SCHOLASTIC

GCSE 9–1
CHEMISTRY
AQA REVISION GUIDE

Mike Wooster

Author Mike Wooster
Editorial team Haremi Ltd
Series designers emc design ltd
Typesetting Newgen KnowledgeWorks (P) Ltd, Chennai, India
Illustrations York Publishing Services and Newgen KnowledgeWorks (P) Ltd, Chennai, India
App development Hannah Barnett, Phil Crothers and Haremi Ltd

Designed using Adobe InDesign
Published by Scholastic Education, an imprint of Scholastic Ltd, Book End, Range Road, Witney, Oxfordshire, OX29 0YD
Registered office: Westfield Road, Southam, Warwickshire CV47 0RA
www.scholastic.co.uk

Printed by Bell & Bain Ltd, Glasgow
© 2017 Scholastic Ltd
1 2 3 4 5 6 7 8 9   7 8 9 0 1 2 3 4 5 6

British Library Cataloguing-in-Publication Data
A catalogue record for this book is available from the British Library.
ISBN 978-1407-17678-9

Acknowledgements

The publishers gratefully acknowledge permission to reproduce the following copyright material:

p9 Hurst Photo/Shutterstock; p11 bszef/Shutterstock; p17 Djordje Konstantinovic/Shutterstock; p20 Daxiao Productions/Shutterstock; p23 Andraž Cerar/Shutterstock; p31 nikkytok/Shutterstock; p35 chromatos/Shutterstock; p36 108MotionBG/Shutterstock; p39 top Billion Photos/Shutterstock; p39 bottom Fablok/Shutterstock; p41 Andrey Kucheruk/Shutterstock; p48 alice-photo/Shutterstock; p50 honglouwawa/Shutterstock; p55 Smith1972/Shutterstock; p62 Natalia Evstigneeva/Shutterstock; p64 left My name is boy/Shutterstock; p64 right Kucher Sergey/Shutterstock; p74 Stocksnapper/Shutterstock; p79 joker1991/Shutterstock; p94 Alexey Stiop/Shutterstock; p104 Perry Harmon/Shutterstock; p109 Pix One/Shutterstock; p115 Aun Photographer/Shutterstock; p119 PhotostockAR/Shutterstock; p120 Saroj Khuendee/Shutterstock; p122 PK289/Shutterstock; p127 PR Image Factory/Shutterstock; p134 exopixel/Shutterstock; p138 Italianvideophotoagency/Shutterstock; p147 steveball/Shutterstock; p148 kunmanop/Shutterstock; p150 PhotoStock10/Shutterstock; p158 Cory Seamer/Shutterstock; p159 Arjen Dijk/Shutterstock.

Every effort has been made to trace copyright holders for the works reproduced in this book, and the publishers apologise for any inadvertent omissions.

Note from the publisher:

Please use this product in conjunction with the official specification and sample assessment materials. Ask your teacher if you are unsure where to find them.

Contents

3

Contents

Topic 6
Topic 7
Topic 8
Topic 9
Topic 10

4

How to use this book

This Revision Guide has been produced to help you revise for your 9–1 GCSE in AQA Chemistry. Broken down into topics and subtopics it presents the information in a manageable format. Written by a subject expert to match the new specification, it revises all the content you need to know before you sit your exams.

The best way to retain information is to take an active approach to revision. Don't just read the information you need to remember – do something with it! Transforming information from one form into another and applying your knowledge through lots of practice will ensure that it really sinks in. Throughout this book you'll find lots of features that will make your revision an active, successful process.

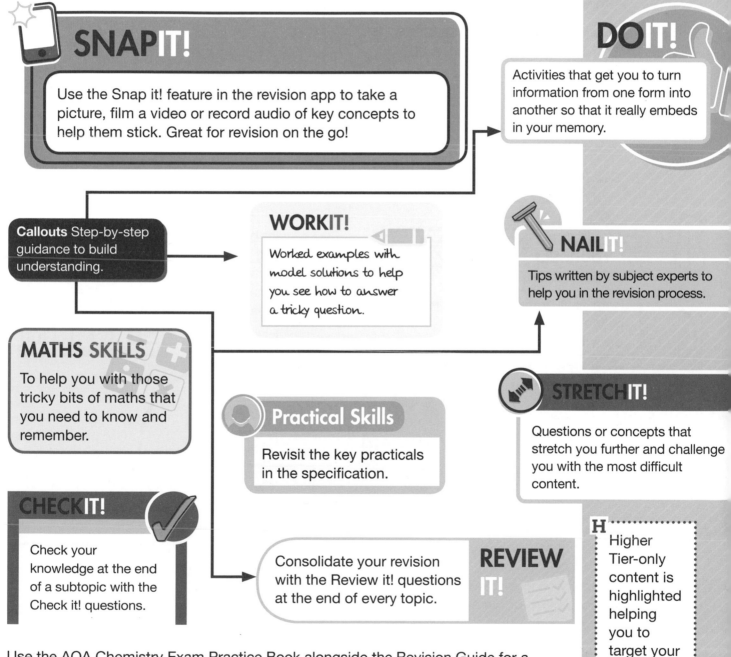

SNAPIT!

Use the Snap it! feature in the revision app to take a picture, film a video or record audio of key concepts to help them stick. Great for revision on the go!

DOIT!

Activities that get you to turn information from one form into another so that it really embeds in your memory.

Callouts Step-by-step guidance to build understanding.

WORKIT!

Worked examples with model solutions to help you see how to answer a tricky question.

NAILIT!

Tips written by subject experts to help you in the revision process.

MATHS SKILLS

To help you with those tricky bits of maths that you need to know and remember.

Practical Skills

Revisit the key practicals in the specification.

STRETCHIT!

Questions or concepts that stretch you further and challenge you with the most difficult content.

CHECKIT!

Check your knowledge at the end of a subtopic with the Check it! questions.

REVIEWIT!

Consolidate your revision with the Review it! questions at the end of every topic.

H Higher Tier-only content is highlighted helping you to target your revision. **H**

Use the AQA Chemistry Exam Practice Book alongside the Revision Guide for a complete revision and practice solution. Packed full of exam-style questions for each subtopic, along with full practice papers, the Exam Practice Book will get you exam ready!

HOW TO REVISE!

PLAN YOUR REVISION

Get ahead by planning your revision!

Work out the **time** you have available for revising.

Think about when you work at your best. Are you a morning or an evening person?

Allocate **MORE TIME** for the topics you struggle with.

Revision works best in **SMALL BURSTS**, so keep sessions **SHORT AND SWEET**!

Remember to allow time to **PRACTISE** applying what you have revised.

Use your **revision app** to put together a revision timetable.

LOOK AFTER YOURSELF

Help your brain by looking after your whole body!

Take regular **breaks** from revising – your brain needs time to digest information in order to retain it.

HOTEL

Keep **hydrated** by drinking plenty of water – dehydration stops your brain from working at its full capacity.

Regular **exercise** helps stimulate the brain and will help you relax.

Get plenty of **sleep**, especially the night before an exam.

EAT WELL and limit unhealthy snacks – your brain needs fuel for memory and concentration.

Find methods of **relaxation** that work for you throughout the revision period.

BE PREPARED!

Limit potential stress on the day of an exam by getting everything you need ready the night before.

30

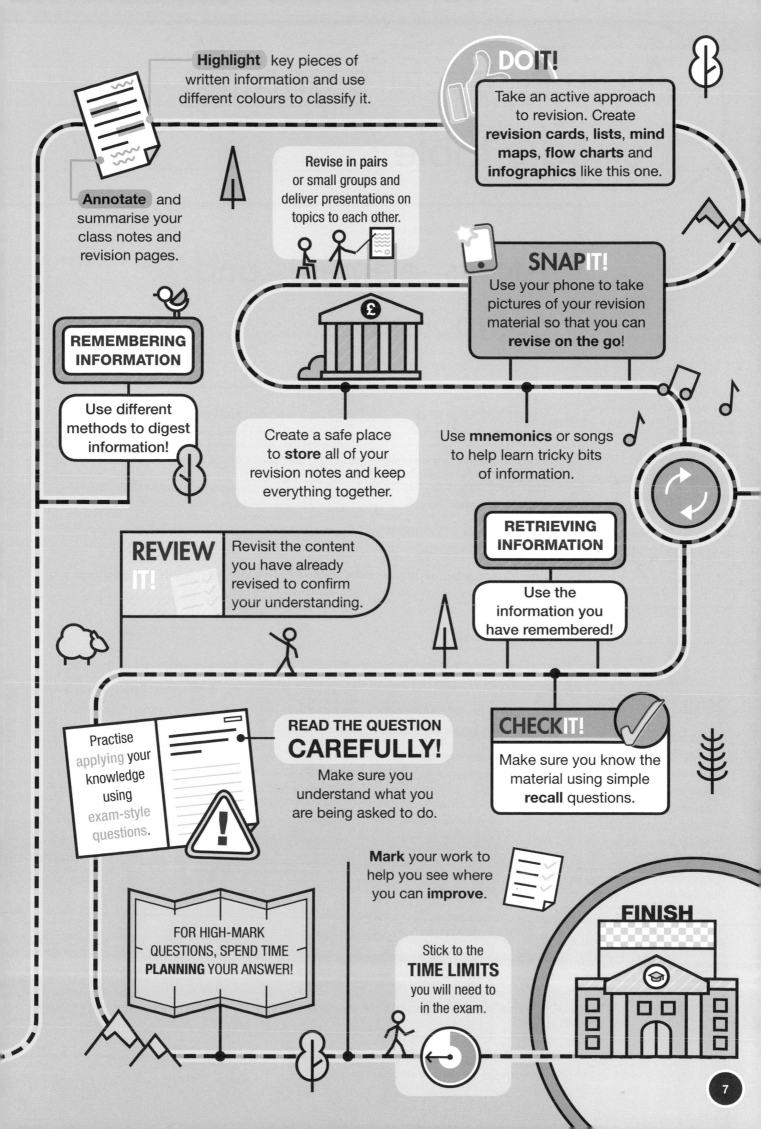

Highlight key pieces of written information and use different colours to classify it.

DO IT!
Take an active approach to revision. Create **revision cards**, **lists**, **mind maps**, **flow charts** and **infographics** like this one.

Annotate and summarise your class notes and revision pages.

Revise in pairs or small groups and deliver presentations on topics to each other.

SNAP IT!
Use your phone to take pictures of your revision material so that you can **revise on the go!**

REMEMBERING INFORMATION

Use different methods to digest information!

Create a safe place to **store** all of your revision notes and keep everything together.

Use **mnemonics** or songs to help learn tricky bits of information.

RETRIEVING INFORMATION

Use the information you have remembered!

REVIEW IT!
Revisit the content you have already revised to confirm your understanding.

Practise applying your knowledge using exam-style questions.

READ THE QUESTION CAREFULLY!
Make sure you understand what you are being asked to do.

CHECK IT!
Make sure you know the material using simple **recall** questions.

Mark your work to help you see where you can **improve**.

FOR HIGH-MARK QUESTIONS, SPEND TIME **PLANNING** YOUR ANSWER!

Stick to the **TIME LIMITS** you will need to in the exam.

FINISH

7

Atomic structure and the periodic table

Atoms, elements and compounds

All substances are made up of particles.

The atom is the smallest particle of an element that can take part in a chemical reaction.

There are over 100 elements and each element contains only one type of atom.

Each element is represented by its own chemical symbol and has its own atomic number.

A compound is formed when the atoms of two or more elements are joined by chemical bonds.

Chemical formulae are used to show which elements are in a compound and the number of atoms of each element that is present.

WORKIT!

The formula of iron(II) sulfate is $FeSO_4$. Identify the elements present in this compound and determine the number of atoms of each element.

Note: The roman numerals are not the number of atoms but the charge on the ion present in the compound.

The elements present are iron, sulfur and oxygen. There is 1 atom of iron, 1 atom of sulfur and 4 atoms of oxygen.

CHECKIT!

1 What type of atom would you find in the element sodium?

2 Explain why the chemical formula H_2 represents an element.

3 What type of substance is silver oxide?

4 What type of substance is represented by the following chemical formulae?

   a $CuBr_2$ b Mg

5 Name the elements present in the following compounds and give the number of atoms of each element.

   a $AgNO_3$ b $Fe(NO_3)_3$

Mixtures and compounds

A physical property of a substance is one which you can measure or observe without changing the substance. Examples of physical properties are melting point and appearance.

A chemical property of a substance is one which you can only observe by changing the substance.

A mixture is formed whenever elements or compounds are together but **not** joined chemically. An example of a mixture is iron mixed with sulfur.

Because the substances in a mixture are not joined chemically they can be easily separated. For example, the iron and sulfur can be easily separated using a magnet.

Mixtures have different physical properties to the substances that make them up.

Because they are separate the substances in a mixture still keep their chemical properties.

There are different types of mixture and this means that there are different separation methods. See the Snap it! box on page 10.

The elements in a compound are joined chemically and therefore cannot be separated by physical means. For example, the compound iron sulfide cannot be separated into iron and sulfur using a magnet because they are combined chemically.

Practical Skills

The compulsory practicals that use these methods are:

1   the preparation of a soluble salt from an insoluble base

2   distillation of salt solution

3   separation of coloured substances using paper chromatography

In some practicals, you will use more than one separation method. For example, in the separation of a soluble salt from an insoluble base you will first use filtration to remove any unreacted insoluble base and then crystallisation to get the salt from the solution.

DO IT!

Start a list of definitions and keep your list on a spreadsheet. You can begin this with definitions of an atom and a compound.

DO IT!

In this table there is an empty column labelled 'Example'.

For each separation method, give an example of where it is used. The answers are given below. Just put each example in the right box.

Ethanol from ethanol and water; salt from salty water; chalk from chalk and water; water from salty water; colourings in sweets.

NAIL IT!

Some of the marks (about 15%) will be allocated to questions about practical techniques. For some of the required practical experiments you use some of these methods. Make sure you know which separation method is used along with the apparatus that is needed.

In the exam you may be given data on the different boiling points or solubilities of different substances and asked to explain how you could separate them.

SNAP IT!

Table to show methods for separating different mixtures

Mixture	Method used and why	Apparatus used	Example
Separating an insoluble solid from a liquid.	Filtration because the insoluble solid cannot pass through the filter paper.	Filter funnel; filter paper and beakers.	
Separating the liquid from a solution of a solid in a liquid. The liquid is the distillate.	Simple distillation because the liquid has a much lower boiling point and so evaporates at a much lower temperature.	Flask; heating equipment and condenser.	
Separating two or more miscible liquids. (miscible means they can mix)	Fractional distillation because the liquid with the higher boiling point condenses on the column as the one with the lower boiling point carries on up as a vapour.	Flask; heating equipment; fractionating column and condenser.	
Separating coloured substances.	Paper chromatography which relies on the substances having different attractions for the paper and the solvent.	Container and chromatography paper.	
Separating the dissolved solid from a solution.	Crystallisation, which depends on the big differences in boiling points between the solvent and the dissolved solid.	Evaporating basin and heating equipment.	

WORKIT!

The boiling points of two substances X and Y along with water are shown in the table below.

Substance	Boiling point/°C
X (X is a solid at room temperature)	1800
Y (Y is a liquid at room temperature)	67
Water	100

a   Explain how you could get X from a solution of X in water.

Crystallisation. The water has a much lower boiling point and can be evaporated off to give solid X.

b   Explain how you could get Y from a solution of Y in water.

Fractional distillation. Y and water are miscible (they mix), otherwise Y would not dissolve in water. Their boiling points are close so fractional distillation is needed.

✓ CHECKIT!

1 Why is a mixture of iron and sulfur easy to separate but it is very difficult to separate iron from sulfur in iron sulfide?

2 a Explain how you could separate a mixture of chalk and salt. Check the Snap it! for ideas.

b The table below shows the solubilities of two solid substances, Q and R, in petrol and water.

Substance	Q	R
Solubility in petrol	Soluble	Insoluble
Solubility in water	Insoluble	Insoluble

Explain how you could use filtration to separate a mixture of Q and R.

Scientific models of the atom

From the ancient Greeks up to the end of the 19th century, atoms were thought to be indivisible.

Joseph John Thomson discovered the electron and he suggested that the atom is a positive ball (the plum pudding) with negatively charged electrons (the currants) dotted around inside it.

A few years later, Ernest Rutherford showed that the atom has a central nucleus which contains positively charged protons with electrons orbiting around it.

It is now thought that the electrons are in energy levels or shells around the nucleus.

James Chadwick discovered the neutron (which is neutral) in the nucleus. This means that there are three sub-atomic particles – the electron, the proton and the neutron.

NAILIT!

The most important experiment is Rutherford's experiment. When Rutherford and his team fired high-energy positively charged alpha particles at gold foil they expected these particles to pass straight through. What they saw was that most of the particles did pass straight through but some were deflected or rebounded straight back.

Think about Rutherford's reasoning on his experimental results. The nucleus must be positive to repel the positive alpha particles and very dense because it had to withstand their high energy. Because most of the alpha particles passed through the foil, most of the atom must be empty space.

DOIT!

Copy and complete the table below to show how scientists' vision of the atomic model atom has changed over time as new evidence became available.

Scientist	What they discovered	Comments

SNAPIT!

The diagram below shows the model of the atom we now use.

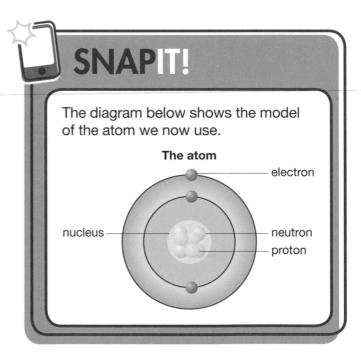

The atom

- electron
- nucleus
- neutron
- proton

CHECKIT!

1 What does indivisible mean?

2 Name the three main sub-atomic particles.

3 Using the names of these three particles describe the structure of the atom.

Atomic structure

In an atom, the protons and neutrons are in the central nucleus and the electrons move around the nucleus in electron shells.

The charges and masses of the three sub-atomic particles are very small and because of this we use their relative charges and relative masses.

See the table below for a summary of their properties.

Each of the atoms of an element contains the same number of protons and this is called the atomic number.

For each element the atomic number is fixed and cannot change.

In a neutral atom the number of protons equals the number of electrons.

Ions are charged atoms and are formed when atoms of an element react with atoms from another element.

An ion is positive if electrons are lost and the number of positive charges on the ion is equal to the number of electrons lost.

An ion is negative if electrons are gained and the number of negative charges is equal to the number of electrons gained.

The mass number of an atom is the sum of the number of protons and neutrons in the nucleus.

Table showing properties of sub-atomic particles

Name of sub-atomic particle	Where is it in the atom?	Relative charge	Relative mass
Proton	In the nucleus	+1	1
Neutron		0	1
Electron	In shells around the nucleus	−1	Very small

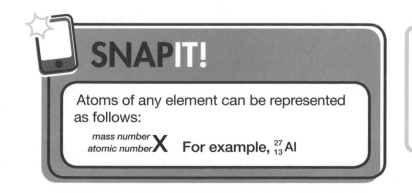

SNAPIT!

Atoms of any element can be represented as follows:

$_{atomic\ number}^{mass\ number} X$ For example, $_{13}^{27}Al$

DOIT!

Draw a diagram of an atom, label each sub-atomic particle in your diagram and write a short note to describe its relative mass and its relative charge.

Use the words electron, proton, neutron, atomic number, mass number, nucleus and neutral in your notes.

NAIL IT!

You must remember that the number of protons (the atomic number) in atoms of the same element never changes. If its atomic number were to change, it would be a different element.

MATHS SKILLS

You can work out the number of each sub-atomic particle in an atom by using its atomic number and mass number.

In a **neutral atom** the atomic number = number of protons = number of electrons

The mass number = number of protons + number of neutrons = atomic number + number of neutrons

This means that: the number of neutrons = mass number − atomic number

In a **positive ion** the number of electrons = atomic number − the number of charges on the ion

In a **positive ion** the number of electrons = atomic number + the number of charges on the ion

WORKIT!

1 An atom of potassium has the atomic number 19 and mass number 39. What are the number of electrons, protons and neutrons in an atom of potassium?

The number of protons and electrons are equal to the atomic number = 19

The number of neutrons = mass number − atomic number = 39 − 19 = 20

2 How many electrons are there in a calcium ion Ca^{2+}? [Atomic number of calcium = 20]

For Ca^{2+} number of electrons = 20 − 2 = 18

3 How many electrons are there in N^{3-} ion? [Atomic number of nitrogen = 7]

For N^{3-} number of electrons = 7 + 3 = 10

CHECK IT!

1 An oxygen atom contains 8 protons. Why can't it have 9 protons?

2 Explain why virtually all of the mass of the atom is found in the nucleus.

3 A phosphorus atom (symbol P) has an atomic number 15 and a mass number 31. Show how you could represent the phosphorus atom.

4 How many electrons are there in an Al^{3+} ion? [Atomic number of aluminium = 13]

5 An atom of sodium can be represented as shown below. Give the number of protons, electrons and neutrons in a sodium atom.

$^{23}_{11}Na$

Isotopes and relative atomic mass

The atomic number of an element cannot change. It is fixed.

Therefore the number of protons is also fixed.

But the mass number of an element can have different values and so the number of neutrons must also vary in number.

Atoms of the same element with different mass numbers are called isotopes.

Isotopes of an element have the same number of protons but different numbers of neutrons.

Examples of isotopes are the three naturally occurring isotopes of magnesium:

a $^{24}_{12}Mg$ **b** $^{25}_{12}MG$ **c** $^{26}_{12}Mg$

These all have 12 protons in the nucleus but atom a) has 12 neutrons, b) has 13 neutrons and c) has 14 neutrons in the nucleus.

For any element there are different amounts of each isotope and this has to be taken into account when calculating the relative atomic mass of an element.

The units of atomic mass are atomic mass units (amu).

MATHS SKILLS

The amount of an isotope in terms of its percentage is its percentage abundance.

When you work out the relative atomic mass of an element you start off by saying 'Let there be 100 atoms'. The percentage abundance of an isotope is the number of atoms out of the 100 which are that isotope.

You then multiply the percentage abundance of the isotope by its mass number to give the mass due to its atoms.

Repeat this for the other isotopes and add up all the masses due to all the isotopes.

Finally divide the total mass by 100 to get the average mass which is the relative atomic mass.

NAIL IT!

The relative atomic mass is a weighted average which means that we don't just add up the mass numbers of the isotopes and find the average. We have to take into account the abundance of each isotope. If you consider the two isotopes of chlorine $^{35}_{17}Cl$ and $^{37}_{17}Cl$, the average is 35+37/2 =36 but this does not take into account that there is more of the $^{35}_{17}Cl$ isotope and that is why the relative atomic mass is 35.5amu – nearer to 35.5. This means that when working out the relative atomic mass its value should be near the most abundant isotope.

DO IT!

The mass numbers for the isotopes of chromium are shown below. Estimate the relative atomic mass of chromium to the nearest whole number and then look up the relative atomic mass in the periodic table.

Element	Mass number for each isotope with percentage abundance in brackets
Chromium	50 (4.31%); 52 (83.76%); 53 (9.55%) and 54 (2.36%)

WORKIT!

1 An example is chlorine with its two isotopes $_{17}^{35}Cl$ and $_{17}^{37}Cl$. The isotope $_{17}^{35}Cl$ makes up 75% of the atoms and the $_{17}^{37}Cl$ isotope 25%. Calculate the relative atomic mass of chlorine.

Let there be 100 atoms. 75 of these are the $_{17}^{35}Cl$ isotope and they have a total mass of 75 × 35 = 2625amu

The total mass of the $_{17}^{37}Cl$ isotope = 25 × 37 = 925amu

The total mass of 100 atoms of all the isotopes = 2625 + 925 amu = 3550amu

The average mass = 3550/100 = 35.5amu = the relative atomic mass of chlorine.

2 There are three naturally occurring isotopes of magnesium $_{12}^{24}Mg$ (78.6% of total); $_{12}^{25}Mg$ (10.11% of total) and $_{12}^{26}Mg$ (11.29% of total). What is the relative atomic mass of magnesium?

Even though the percentages are not whole numbers just use the same method as for chlorine. One way of checking your answer is to estimate which number the relative atomic mass would be nearest to. In this case the most abundant isotope is magnesium-24 and therefore you expect the relative atomic mass would be nearer to 24 than the others.

Let there be 100 atoms. Mass of magnesium—24 isotope = 78.6 × 24 amu = 1886.4 amu

Mass of magnesium-25 isotope = 10.11 × 25 amu = 252.8 amu

Mass of magnesium-26 isotope = 11.29 × 26 amu = 293.5 amu

The total mass of 100 atoms = 2432.7 amu

Therefore the relative atomic mass is the average mass of each atom = 2432.7/100 = 24.3 amu (to 3 significant figures).

As expected, this is near to 24 which is the mass number of the most abundant isotope.

CHECKIT! ✓

1 a Explain why the two atoms represented by $_{18}^{38}Ar$ and $_{18}^{40}Ar$ are isotopes

b Give the numbers of electrons, protons and neutrons in these two atoms.

2 a There are two naturally occurring isotopes of copper. Their mass numbers and percentage abundance are given in this table.

Mass number	Percentage abundance
63	69%
65	31%

Use this data to calculate the relative atomic mass of copper to 3 significant figures.

b The atomic number of copper is 29. How many electrons, protons and neutrons are there in each isotope of copper?

The development of the periodic table and the noble gases

The periodic table we have today was developed by Mendeleev.

At the time, Mendeleev organised the elements in order of their atomic weights and he placed elements with similar properties in groups.

He did not place elements where they did not fit and left spaces for elements that he thought had not yet been discovered.

He also predicted the properties of these missing elements. When they were discovered his predictions were found to be very accurate.

Today the elements are placed in order of their atomic number. If the elements were placed in order of their relative atomic mass the elements iodine and tellurium would be placed in the wrong groups. The same applies to argon and potassium.

Vertical columns of elements are called groups and horizontal rows of elements are called periods.

The number of electrons in the outer shell of an element is its group number.

The number of occupied electron shells is the period number.

The first of the noble gases to be discovered was argon. This new element had completely different properties to those elements already discovered.

Because of Mendeleev's concept of groups of elements with similar properties, argon's discoverers then started looking for other elements with similar properties. This led to the discovery of the other noble gases: helium (in the Sun) and neon, krypton and xenon from the fractional distillation of liquid air.

The noble gases are very unreactive because they have full outer electron shells which are stable electron arrangements.

As you go down the group the noble gases become denser.

DO IT!

Write a short description of the present-day periodic table using the words below:

groups
similar
period
columns
electron shells
properties
atomic number
horizontal rows

SNAPIT!

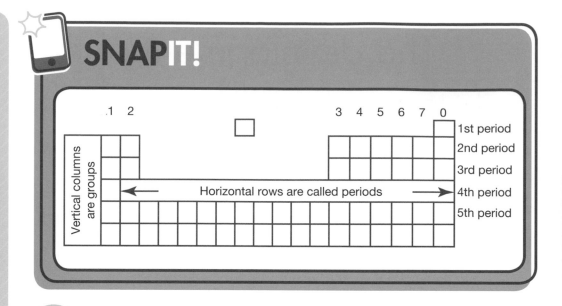

NAILIT!

- Once you have located an element in the periodic table you will probably need to select information about the element.
- Remember the top number above the symbol is the relative atomic mass and the number below the symbol is the atomic number.
- The first row of elements consists of hydrogen and helium. These are easily missed because the hydrogen is placed on its own because it does not fit into a particular group.
- The groups you should concentrate on are the ones featured in this book. These are Group 1– the alkali metals, Group 7 – the halogens and Group 0 – the noble gases.

The Periodic Table of the Elements

Key
relative atomic mass
symbol
name
atomic (proton) number

Elements with atomic numbers 112–118 have been reported but not fully authenticated

*58–71 Lanthanide
*90–103 Actinide

CHECKIT!

1 What are the names given to vertical columns and horizontal rows in the periodic table?

2 Name the element that is in Group 3 and Period 3 of the periodic table.

3 Why did Mendeleev leave spaces in his version of the periodic table?

4 Look at the pairs of elements potassium and argon and tellurium and iodine. Use these examples to explain why we do not arrange the elements in order of their atomic mass.

Electronic structure and the periodic table

The electrons in atoms are arranged in electron shells.

The first shell can take a maximum of 2 electrons, the second shell holds 8 electrons and the third shell can also hold 8 electrons.

The number of electrons in the outer shell of an element is its group number.

The number of occupied electron shells is the period number.

The electronic structures with **full outer shells** are particularly important because they are very stable. These structures are the electronic structures of the noble gases – 2, 2, 8 and 2, 8, 8.

SNAPIT!

The diagram below gives two ways to represent the electronic structure of an element. In this case the elements carbon and chlorine.

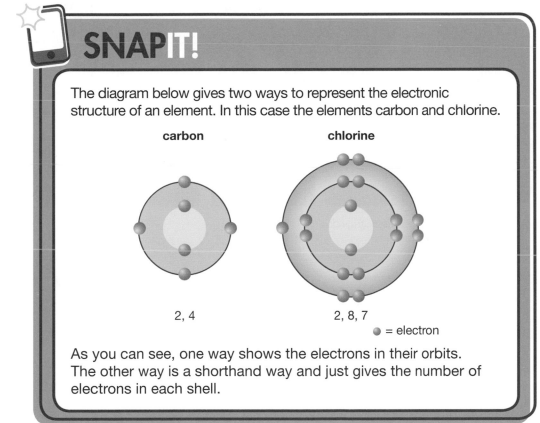

carbon	chlorine
2, 4	2, 8, 7

● = electron

As you can see, one way shows the electrons in their orbits. The other way is a shorthand way and just gives the number of electrons in each shell.

DOIT!

Suppose you were given the atomic number of an element. Describe to a revision partner two ways you could locate an element in the periodic table. (Hint: number of protons and electrons.)

Or record a short MP3 of your explanation and check its accuracy by playing it back and comparing it with your notes or your textbook.

NAILIT!

Electronic structure is very important.

If you can work out the electronic structures of different atoms you can then go on to work out the type of bond formed between them.

The electronic structures with full outer shells are particularly important because they are stable arrangements. These are sometimes called the noble gas arrangements.

You only need to work out the electronic structures of the first 20 elements i.e. those for the elements hydrogen to calcium.

MATHS SKILLS

You will need to work out the electronic structure of an element. You do this by filling the shells up starting from the first or innermost shell which takes up to 2 electrons. You then fill up the second shell until it is full and so on.

WORKIT!

Phosphorus has 15 electrons. Only 2 electrons go into the first shell, leaving 13 electrons. Only 8 can go into the second shell leaving 5 electrons for the outermost shell which can take up to 8.

Write out the electronic structure of phosphorus.

The electronic structure or electron arrangement of phosphorus is written as 2,8,5.

If we use the periodic table we can check this is right by looking at the third period (third row) and Group 5. If we look here we find phosphorus.

Match heads contain phosphorous

CHECKIT!

1 Where are the electrons found in an atom?

2 What is the maximum number of electrons that could fit into the second shell?

3 An element has 19 electrons. What is its electron arrangement and where would you place it in the periodic table?

Metals and non-metals

Metals are found on the left-hand side of the periodic table and non-metals on the right-hand side.

Most elements (about 92 out of 118) are metals.

The two types of element are different in their appearance, electrical conductivity, malleability and ductility. These differences are shown in the table below.

When metals react with non-metals their atoms lose electrons to form positive ions.

When non-metals react with metals their atoms gain electrons to form negative ions.

Physical property	Metals	Non-metals
Electrical conductivity	Electrical conductors	Electrical insulators
Malleability and ductility	Malleable and ductile	The solids are brittle. They snap when you try to bend or stretch them.
Appearance	All shiny	The solids are dull in appearance.

Table to show differences between the physical properties of metals and non-metals

DOIT!

Choose a metal and a non-metal from the periodic table. Look up the properties of both elements on the Internet and check to see if they have all the physical properties listed in the table and record if they form a positive or negative ion.

Don't forget to add the definitions we have used here to your spreadsheet of definitions recommended at the beginning of the guide.

SNAPIT!

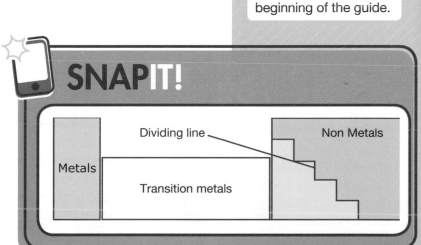

CHECKIT!

1 Calcium is a metal. Where would you find calcium in the periodic table?

2 Classify the elements sulfur and sodium as metals or non-metals. You should explain your answer by referring to some of their physical properties.

3 What type of ion, positive or negative, is formed by the atoms of sodium? Explain how you arrived at your answer.

4 Black phosphorus is a form of phosphorus that conducts electricity. By referring to the periodic table explain why this is unusual.

NAILIT!

Malleability means it can be hammered into shape and ductility means it can be drawn into wires.

Learn the differences in physical properties. If you learn the properties of metals then the properties of non-metals are usually the opposite.

Group 1 – the alkali metals

The alkali metals are all in Group 1 because they have one electron in their outer electron shell.

They are typical metals because they are shiny and good electrical conductors.

But, unlike most other metals, they are soft and have low densities. The first three, lithium, sodium and potassium, are less dense than water.

They all react with water to give hydrogen gas and an alkaline solution of the metal hydroxide. For example, sodium gives hydrogen and sodium hydroxide when it reacts with water.

sodium(s) + water(l) → hydrogen(g) + sodium hydroxide(aq)

$2Na(s) + 2H_2O(l) → H_2(g) + 2NaOH(aq)$

When they react they all lose their outer electron to form a +1 ion (e.g. Na^+). This gives them a stable full outer shell of electrons.

They get more reactive as you go down the group because it gets easier to lose their outer electron. See the table below.

DOIT!

You can find YouTube videos of the reactions of the alkali metals on the Internet. They will show what happens when they are added to water. Just do a search for 'alkali metals reactions with water'.

NAILIT!

There are often questions on the Group 1 elements in the periodic table. You need to know the word and symbol equations for their reactions with water.

You should also be able to predict the reactions of rubidium and caesium using what you know about the first three elements in the group.

SNAPIT!

Li	Hardest in group but can be cut with a knife, tarnishes (goes dull) in air. When it is added to water it floats, fizzes because hydrogen is released and leaves an alkaline solution of lithium hydroxide (LiOH).	They all have 1 electron in their outer electron shell and this is why they are in Group 1.
Na	Easier to cut with a knife than lithium, tarnishes quickly. When added to water it moves around on the surface of the water and gets hot enough to melt, giving hydrogen gas and alkaline sodium hydroxide (NaOH) solution.	They are all soft and easily cut with a knife. They all react with air and water and this is why they are stored under oil.
K	Even softer and tarnishes immediately. With water it whizzes around the surface and melts, and the hydrogen given off burns producing a lilac flame. The solution remaining contains potassium hydroxide (KOH).	As you go down the group their reactions with water get more and more violent. Each reaction produces hydrogen and an alkaline solution of the metal hydroxide.
Rb	Look up Internet videos.	
Cs		

WORKIT!

What is the electron arrangement of sodium?

Sodium has the atomic number 11. This means that it has 11 electrons. 2 electrons go in the first shell, 8 in the second and this leaves 1 for the outer third shell. Its electronic structure is 2,8,1.

MATHS SKILLS

You should be able to work out the electron arrangements of the first three alkali metals. Remember 2 electrons can go in the first shell, 8 in the second and 8 in the third.

A piece of sodium

NAILIT!

Make sure you can explain why the elements get more reactive as you go down the group:

- They all lose their outer single electron when they react to give a stable electron arrangement.

- This gets easier as you go down the group for two reasons.

- The outer electron gets further from the positive nucleus and so it feels less of an attractive force and can leave the atom more easily.

- As you go down the group there are more electron shells between the nucleus and the outer electron. This also lowers the attractive force from the nucleus.

✓ CHECKIT!

1 a Give the electron arrangements of lithium (atomic number 3) and potassium (atomic number 11).

b Use your answers to explain why they are in the same group of the periodic table.

2 Give four physical properties of sodium.

3 Give the word and balanced symbol equations for the reaction of potassium with water.

4 Why is sodium more reactive than lithium?

5 a Predict the observations you would make if rubidium was added to water.

b Write the balanced symbol equation for the reaction of rubidium with water.

6 Explain why the alkali metals are always found in compounds never uncombined.

Group 7 – the halogens

The **halogens** are all in Group 7 because they have 7 electrons in their outer electron shell.

As elements they exist as **molecules** of 2 atoms (**diatomic molecules**) e.g. Cl_2 and Br_2.

When they react with metals they all gain 1 electron to form a −1 ion. These ions are called **halide** ions.

As you go down the group the:

- elements get heavier as the relative molecular masses increase.

- elements get **less reactive** because it gets harder to gain an extra electron (see Snap it! box).

- more reactive halogens **displace** less reactive ones from solutions of their salts (see next sub-topic).

DOIT!

Using the Snap it! box, describe the properties of the elements as you go down the group. Predict the properties of astatine and then look it up on the Internet. How many did you get right?

SNAPIT!

Pale yellow gas	F	They all have 7 electrons in their outer shell and this is why they are in Group 7.
Pale green gas	Cl	They get darker as you go down the group and their melting points and boiling points increase.
Dark red liquid	Br	They all exist as molecules of 2 atoms and form -1 ions when they react with metals.
Dark grey solid	I	They get **less reactive** as you go down the group.
Black solid	At	This is because it gets harder to gain an extra electron to become stable.

NAILIT!

There are often questions on the Group 7 elements in the periodic table.

You should also be able to predict the reactions of fluorine and astatine using what you know about the middle three elements in the group: chlorine, bromine and iodine.

Of the Group 7 elements you will only be asked to work out the electron arrangements (or draw the electron structure) for fluorine and chlorine. However, you should be able to work out that the electron arrangement of iodine and bromine also end in 7.

One common error is to use chloride instead of chlorine. Chloride is used when chlorine is part of a compound. You can have sodium chloride but not sodium chlorine. The element 'chloride' does not exist.

WORKIT!

Bromine is in Group 7. It has two isotopes $^{79}_{35}Br$ (50.5%) and $^{81}_{35}Br$ (49.5%).

a   Calculate the relative atomic mass of bromine.

The first step to remember is to let there be 100 atoms and then work out the mass of each isotope in the 100 atoms.

Mass of bromine – 79 isotopes = 50.5 × 79 = 3990 amu.

Mass of bromine – 81 isotopes = 49.5 × 81 = 4010 amu

The total mass of 100 atoms = 3990 + 4010 = 8000 amu

This makes the relative atomic mass = 80

b   Calculate the number of electrons, protons and neutrons in the bromine-79 isotope.

The atomic number is 35 and this means that there must be 35 electrons and 35 protons.

The mass number = 79 and the number of neutrons = 79 – 35 = 44

c   How many electrons are there in a Br^- ion?

Because the ion is 1– it must have 1 extra electron compared with the neutral atom. The number of electrons = 35 + 1 = 36 electrons.

MATHS SKILLS

Exam questions often need you to use different techniques and areas of knowledge and these link together different areas of the specification. For example, in the same question you may need to work out relative atomic mass and numbers of sub-atomic particles.

NAILIT!

Make sure you can explain why the elements get less reactive as you go down the group:

- They all gain an extra electron when they react to give a stable electron arrangement.

- This gets harder as you go down the group for two reasons.

- The outer electron shell gets further from the positive nucleus and so any electron feels less of an attractive force and this makes it harder to gain an electron.

- As you go down the group there are more electrons between the nucleus and the outer electron. This also lowers the attractive force from the nucleus and makes it harder to gain an electron.

CHECKIT!

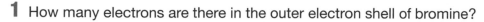

1   How many electrons are there in the outer electron shell of bromine?

2   When chlorine reacts with sodium what ion is formed, Cl^+ or Cl^-?

3   Draw the electron arrangement of a chloride ion.

Displacement reactions in Group 7

A displacement **reaction** is a chemical **reaction** in which a more reactive element displaces a less reactive element from its compound.

In Group 7 a displacement reaction takes place when a more reactive halogen takes the place of another less reactive halogen in a compound.

A halide is a compound formed between a halogen and another element.

The more reactive halogen displaces a less reactive halogen ion in a solution of a metal halide (a salt).

For example, when bromine is added to a solution of sodium iodide, the more reactive bromine displaces the iodide ion to form sodium bromide and iodine.

This is because bromine accepts an electron more easily than iodine so the iodide ion donates its extra electron to a bromine atom.

DO IT!

Write word equations using the results table in the Snap it! box and use these to explain the observations for each reaction. Then give the order of reactivity from your observations.

SNAP IT!

The results table below shows what happens in some of the halogen displacement reactions. Can you work out the order of reactivity? Check your answer online to see if you were right.

Halogen	sodium chloride solution	sodium bromide solution	sodium iodide solution
chlorine	X	turns yellow/ pale orange	turns brown
bromine	No change	X	turns brown
iodine	No change	No change	X

In solution, chlorine is very pale green; bromine is pale yellow to orange and iodine is brown or purple

NOTE: X is placed where there is no experiment. For example, you cannot displace a chloride using chlorine!

MATHS SKILLS

You should be able to balance the equations for displacement.

First, write down the word equation, then the symbol equation and then balance it.

WORK IT!

The word equation for reacting sodium bromide with chlorine is:
Sodium bromide + chlorine → sodium chloride + bromine

Write a balanced equation for this reaction.

Write symbols $NaBr + Cl_2 \rightarrow NaCl + Br_2$

Balance equation $2NaBr + Cl_2 \rightarrow 2NaCl + Br_2$

Practical Skills

When you do these experiments you must always add the halogen to water as a control. This means you can compare it with the reaction mixture to see if there is a change.

Sometimes cyclohexane is added to the mixture and shaken. Any halogen formed dissolves more in the cyclohexane than the water. This helps because in water bromine and iodine are not very different in colour depending on their concentrations. In cyclohexane bromine is still orange/red in colour whilst iodine is purple. So this can be used to confirm the identity of the halogen produced.

STRETCHIT!

For the Higher Tier you can be asked to write ionic equations for the displacement reactions that take place. The simple rule for these is that the salts exist as ions and the halogen molecules do not. For example, the reaction between chlorine and potassium iodide solution gives iodine and potassium chloride solution.

Symbol equation: $2KI + Cl_2 \rightarrow 2KCl + I_2$

The potassium iodide and potassium chloride can be written as ions:

$2K^+ + 2I^- + Cl_2 \rightarrow 2K^+ + 2Cl^- + I_2$

The K^+ ions do not change and can be cancelled out. Because it does not take part in the reaction the K^+ ion is called a **spectator ion** (it just looks on).

So the final equation is $2I^- + Cl_2 \rightarrow 2Cl^- + I_2$

For more information on ions, see page 33.

CHECKIT!

1 What is the formula of a fluorine molecule?

2 What would you see happen if chlorine was added to a solution of sodium iodide?

3 a Give the word and symbol equations for the reaction between bromine and sodium iodide (NaI).

  b Describe what you would see happen when the bromine is added to the sodium iodide. Explain this observation.

H c Write the ionic equation for this reaction.

4 When bromine is added to a solution containing sodium chloride no change is observed.

  a What would be the control for this reaction?

  b Explain the observation.

NAILIT!

When you write the formulae for the halogens remember they are diatomic molecules, X_2. So, for example, chlorine is Cl_2 and bromine is Br_2. The general formula of all the sodium halides is NaX.

The transition metals

DOIT!

Draw a rough outline of the periodic table. On it mark where you would find the alkali metals, the halogens, the noble gases and the transition elements.

Write a short list of the properties of each group on sticky notes or on your outline.

NAILIT!

For transition metals concentrate on their coloured compounds and that they each form more than one positive ion.

The transition metals are found in the middle block of the periodic table.

They have typical metal properties. They are shiny, malleable and ductile. They are also good electrical and thermal conductors.

Unlike the Group 1 metals, transition metals are hard and dense.

They are not as reactive as the Group 1 metals. Their reactions with oxygen, water and chlorine are not as vigorous as the reactions of the Group 1 metals.

As for all metals, the atoms of transition metals lose electrons when they react to form positive ions. The difference is that each transition metal forms ions with different charges. For example, iron forms Fe^{2+} and Fe^{3+} ions.

The different metal ions of each transition metal have different colours. For example, Fe^{2+} ions are green and Fe^{3+} ions are orange-brown.

This means that transition metals form coloured compounds. Non-transition metal compounds are all white in colour.

Transition metals and their compounds form good catalysts. Examples are iron in the Haber process, platinum in the catalytic converter in car exhausts and vanadium pentoxide in the manufacture of sulfuric acid.

SNAPIT!

There are only a few coloured compounds that you need to know about.

Colours of compounds containing transition metal ions.

Cu^{2+}	Copper(II)
Fe^{2+}	Iron(II)
Fe^{3+}	Iron(III)

Later on in the course you may be asked to identify the transition metal ion in a compound and you will use the colours in the Snap it! box to confirm they are present when sodium hydroxide is added to a solution of their ions.

CHECKIT!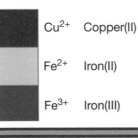

1 Predict four properties of the transition metal chromium.

2 Classify calcium, nickel and iron as either transition or non-transition metals.

3 You have been given unlabelled samples of copper chloride and sodium chloride. Describe how you could tell which is which.

r additional questions, visit:
w.scholastic.co.uk/gcse

1   a   The **unbalanced** symbol equation below represents the reaction between sodium and chlorine.

$$Na(s) + Cl_2(g) \rightarrow NaCl(s)$$

Write the balanced symbol

   b   i   How can you tell from the symbols in the equation that this is a reaction between two elements?

      ii   Describe the appearance of each reactant.

      iii   In the product the two types of particle formed are Na^+ and Cl^-. What is the name given to these charged particles?

   c   Explain why sodium chloride solution in water is a mixture.

2   In the periodic table what is a group and what is a period?

3   Sodium has the atomic number 11. Why can't it also have an atomic number of 12?

4   Sulfur has the atomic number 16. In which group and period of the periodic table would you find sulfur?

5   a   A metal X forms a green compound and a blue compound. Explain where you would find X in the periodic table.

   b   Predict four physical properties of the element X.

6   a   When Group 7 elements react with metals they form ions. What is the charge on these ions?

   b   Explain why the Group 7 elements get less reactive as you go down the group.

7   Magnesium has the atomic number 12 but has atoms with three different mass numbers, 24, 25 and 26.

   a   What does the atomic number 12 tell you about an atom of magnesium?

   b   How is it possible for magnesium atoms to have three mass numbers?

   c   Gallium has two isotopes, gallium-69 (60%) and gallium-71(40%). Use these figures to calculate the relative atomic mass of gallium.

8   Describe two things that Mendeleev did to make his version of the periodic table work.

9   Explain why the elements in Group 1 get more reactive as you go down the group.

10   Explain why the noble gases are so unreactive.

11   a   When the element argon was discovered it had very different properties to any other element. Why do you think scientists went on to look for other elements like it?

   b   Why was this element so hard to discover?

12   The element astatine is in Group 7. It is so rare that most of its chemistry has been guessed at from its position below iodine in Group 7.

   a   Why do we think it is black in colour?

   b   The symbol for astatine is At. What is the formula of an astatine molecule?

   c   Predict the formula of sodium astatide.

   d   What is the charge on an astatide ion?

Bonding, structure and the properties of matter

Bonding and structure

Strong attractive forces between the particles in a substance make its melting and boiling point high. This is because more energy is needed to overcome these forces.

You can work out the state of a substance at room temperature or any other temperature using its melting and boiling points (see Work it! box).

When a substance melts, attractive forces between the solid particles are broken as the particles break away from the solid lattice and become liquid particles. The temperature remains constant until all these bonds are broken.

In the same way when a liquid boils the particles overcome the attractive forces between the liquid particles to become a gas. The temperature stays constant at the boiling point until all these attractive forces are overcome.

The state of a substance at room temperature and pressure (RTP) can be shown in a symbol equation by using the state symbol for the substance.

These state symbols are (s) for solids; (l) for liquids and (g) for gases. The extra symbol (aq) is for substances dissolved in water to give an aqueous solution.

DOIT!

Imagine yourself as a particle. Write a short account of what it would feel like if you were in a solid, a liquid and a gas. What would happen to you if you were a particle in the solid and you were heated until the solid melted?

WORKIT!

A substance X melts at 250°C and boils at 890°C. What state is X in at a) 140°C and b) 723°C?

a 140°C is below the melting point of X so it hasn't melted and is a solid.

b 723°C is above the melting point so X would have melted but it is below the boiling point and this means that it has not boiled and at this temperature X is a liquid.

MATHS SKILLS

You could be asked to work out what state a substance is in at a given temperature. You will be given the melting point and boiling point of the substance.

To work out the state you have to answer two questions. These are – 'At this temperature has it melted?' and 'At this temperature has it boiled?'

SNAPIT!

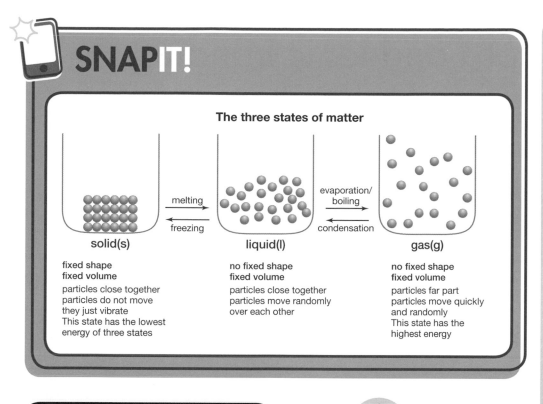

The three states of matter

melting →
← freezing

evaporation/
boiling →
← condensation

solid(s)

fixed shape
fixed volume

particles close together
particles do not move
they just vibrate
This state has the lowest
energy of three states

liquid(l)

no fixed shape
fixed volume

particles close together
particles move randomly
over each other

gas(g)

no fixed shape
fixed volume

particles far part
particles move quickly
and randomly
This state has the
highest energy

NAILIT!

The three states of matter can be represented by a simple model, as shown in the Snap it! box. This particle model can help to explain melting, boiling, freezing and condensing. However for Higher Tier you will be expected to understand the limitations of this model:

• The model assumes that the particles are spheres

• Many particles are not spheres e.g. polymers

• In the model there are no forces between the particles, which is not correct!

CHECKIT!

1 Which state of matter could be described as having a fixed volume but no fixed shape?

2 If a substance has a low melting point what can you say about the forces of attraction between its particles?

3 When a substance melts what happens to the movement and arrangement of the particles?

4 A substance Y has a melting point of −25°C and a boiling point of 135°C.

  a In which state of matter is Y at a) 20°C and b) 250°C?

  b State which state symbol you could use for Y in a symbol equation.

H 5 The particle theory states that particles are solid spheres. Give two reasons why this statement is incorrect.

Ions and ionic bonding

Ions are formed when metal atoms transfer their outer electrons to the outer shells of non-metal atoms. This electron transfer forms charged particles called ions.

Metal atoms form positive ions because they have lost negatively charged electrons. The positive charge on the ion is the same as the number of electrons that are lost.

Non-metal atoms form negative ions because they have gained negatively charged electrons. The negative charge on the ion is the same as the number of electrons that are gained by the atom.

The elements in groups 1 and 2 of the periodic table are metals and they form positive ions.

Each Group 1 atom forms a 1+ ion by losing 1 electron and each Group 2 atom forms a 2+ atom by losing 2 electrons.

The elements in groups 6 and 7 are non-metals and they form negative ions.

Each Group 6 atom gains 2 electrons to form a 2− ion and each Group 7 atom gains 1 electron to form a 1− ion.

All the ions formed have the same electronic structure as the nearest noble gas because these are stable electron arrangements.

The ionic bond is the electrostatic attraction between the positive and negative ions.

You can work out the formula of an ionic compound by balancing the charges on the ions (see Work it! box on the next page).

DO IT!

The atomic number of lithium is 3 and fluorine's is 9. Draw diagrams of the atoms including their nuclei before reaction and diagrams of the ions after reaction. Use your diagrams to explain why the lithium ion is Li+ and the fluoride ion is F−.

SNAP IT!

The **dot-and-cross** diagrams below show the outer electronic structures of some atoms and ions before and after electron transfer.

In the third example each chlorine can accept just 1 electron and the magnesium has to lose 2 electrons and this means that each magnesium has to react with 2 chlorines.

$$Na_x + .\overset{..}{\underset{..}{Cl}}: \longrightarrow \left[Na\right]^+ \quad \left[\overset{..}{\underset{..}{xCl}}:\right]^-$$
2,8,1   2,8,7   2,8   2,8,8

$$Mg^x_x + \overset{..}{\underset{..}{O}}: \longrightarrow \left[Mg\right]^{2+} \quad \left[\overset{..}{\underset{..}{xO}}:\right]^{2-}$$
2,8,2   2,6   2,8   2,8

$$Mg^x_x + 2.\overset{..}{\underset{..}{Cl}}: \longrightarrow \left[Mg\right]^{2+} \quad 2\left[\overset{..}{\underset{..}{xCl}}:\right]^-$$
2,8,2   2,8,7   2,8   2,8,8

NAILIT!

When you draw the ions and their electronic structures you only have to give the outer electrons. For the positive metal ions, you do not have to draw any dots or crosses because they have lost their outer electrons.

You must also remember to draw the electrons gained by the non-metal atom as different to the ones already there. For example, if you look at the chloride ion in the Snap it! box the electron gained from the sodium is written as an × when the chlorine electrons are given as •.

You can also identify whether or not a compound is ionic by looking at the elements. A compound is ionic if one element comes from either Group 1 or 2 and the other one comes from either group 5, 6 or 7.

As far as hydrogen is concerned it resembles Group 7 elements because it has to gain 1 electron to be stable.

WORKIT!

a   Calculate the formula of the compound sodium oxide.

Sodium is in Group 1 which means that its ion is Na^+. Oxygen is in Group 6 which means the oxide ion is O^{2-}.

To make the charges add up to zero we need 2 of the Na^+ ions and 1 of the O^{2-} ion.

The formula is therefore Na_2O.

A dot-and-cross diagram could also be used to find the formula. The sodium atom needs to lose 1 electron and the oxygen needs to gain 2 electrons. Therefore to make this happen there needs to be 2 sodium atoms combining with 1 oxygen and the formula is therefore Na_2O.

b   Calculate the formula of strontium fluoride.

Strontium is in Group 2 and this means that its ion is Sr^{2+}. Fluorine is in Group 7 so its ion is F^-. To make the charges add up to zero we need 1 strontium ion and 2 fluoride ions. The formula is SrF_2.

MATHS SKILLS

When you work out the formula of an ionic compound you have to make the charges on the ions add up to zero. Let's suppose you have a compound containing M^{2+} ions and X^- ions. To make the charges add up to zero we need 2 of the 1− ions and 1 of the 2+ ions. This means that the formula is MX_2.

✓ CHECKIT!

1 The list below shows the formulae of six compounds. From the list choose the three ionic compounds.

   LiCl CS_2 NH_3 $BaBr_2$ CO_2 NaH

2 Draw dot-and-cross diagrams for the three ionic compounds you have chosen.

3 Why do Group 1 elements form 1+ ions?

4 Give the formula of the sulfide ion.

5 Why do the ions in NaCl stay together?

6 What are the formulae of the ionic compounds potassium sulfide and magnesium iodide?

The structure and properties of ionic compounds

The ions in ionic compounds are held together because of the strong electrostatic attraction between the oppositely charged ions.

The electrostatic forces around each ion extend in all directions so each ion attracts several oppositely charged ions around it. This results in a giant lattice where the ions are regularly arranged in a repeating pattern.

The diagrams below showing the structure of sodium chloride illustrate this.

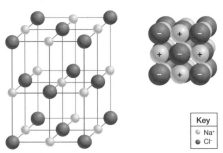

Key
○ Na⁺
● Cl⁻

Even though there are many ions in a giant ionic lattice the numbers of each ion are in the same ratio as they are in the formula of the compound. For example, in NaCl the number of sodium and chloride ions are equal in number. In $CaBr_2$ there are twice as many bromide ions as there are calcium ions.

Ionic compounds have high melting and boiling points because the ionic bonds are **strong** and in the giant lattice there are lots of them to break. You need a lot of energy to break these bonds.

Ions are charged particles and when they move they can carry an electric current.

In solid form ionic compounds do not conduct electricity because the ions are fixed in a lattice and do not move and therefore they cannot carry the current.

If an ionic compound is dissolved in water or is in molten form, then the ions are no longer fixed in a lattice and can move. This means that the ionic compound will conduct electricity.

DO IT!

You may be asked to complete part of an ionic lattice by placing the ions in their places on the grid. The one most usually asked is NaCl because it is 1:1 in terms of positive and negative ions

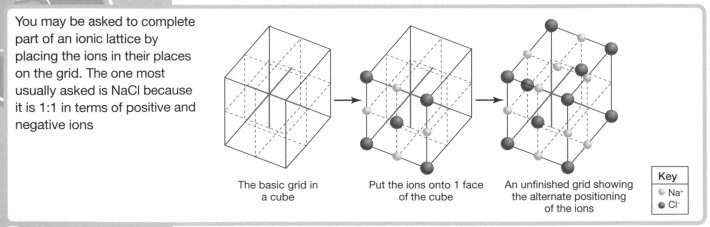

The basic grid in a cube

Put the ions onto 1 face of the cube

An unfinished grid showing the alternate positioning of the ions

Key
○ Na⁺
● Cl⁻

NAILIT!

With a few exceptions all giant structures have high melting and boiling points because they have lots of bonds that have to be broken. The attraction between ions increases as the charges on the ions increase. For example, the attraction between magnesium ions Mg^{2+} and oxide ions O^{2-} is greater than the attraction between sodium ions Na^+ and chloride ions Cl^-. This means that the melting point of magnesium oxide is higher than that of sodium chloride.

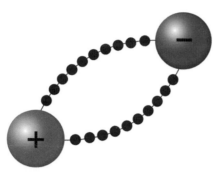

SNAPIT!

Property	Ionic compounds	Explanation
Melting and boiling point	High	Strong electrostatic attraction between the ions and there are lots of bonds to be broken
Electrical conductivity of solid	Poor	The ions cannot move and cannot carry the current
Electrical conductivity of liquid	Good	The ions can move and can carry the current

CHECKIT!

1 What keeps the ions together in an ionic lattice?

2 Explain why the melting point of sodium chloride is very high.

3 Explain why sodium chloride does not conduct electricity as a solid but it does conduct electricity when it is dissolved in water.

4 What can you say about the number of fluoride ions compared to magnesium ions in the ionic lattice of magnesium fluoride (MgF_2)?

5 Explain which of potassium bromide (KBr) and calcium oxide (CaO) has the higher melting point.

Covalent bonds and simple molecules

A covalent bond is formed when a pair of electrons is **shared** between the atoms of two non-metal atoms.

The number of covalent bonds formed by an atom is equal to the number of electrons it needs to gain.

After forming the bonds each atom has the same electronic structure as the nearest noble gas.

Covalent bonds are strong and require a lot of energy to break.

Covalent bonding can be represented by dot-and-cross diagrams or straight lines drawn between atoms. For example:

HCl can be represented by H——Cl or H×Cl

CH_4 can be represented by H——C——H or H C H
(with H above and below the central C)

Covalently bonded substances can exist as simple molecules, giant covalent structures or polymers.

Simple molecules are neutral particles and have no charges or free electrons. Because of this they cannot carry an electric current either in liquid or solid form.

Simple molecular substances have weak intermolecular forces between the molecules. These weak attractive forces need little energy to break them and this means that simple molecular substances have low melting and boiling points and quite often they are liquids or gases at room temperature.

The intermolecular forces increase as the molecular size **increases** and this means that the melting and boiling points also increase.

SNAPIT!

See the examples below:

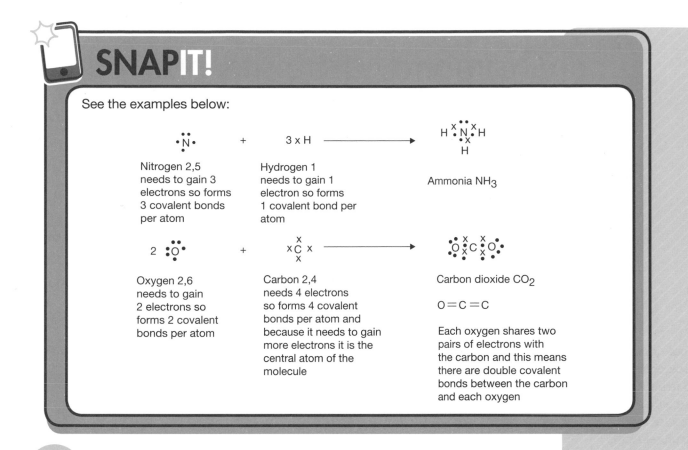

Nitrogen 2,5 needs to gain 3 electrons so forms 3 covalent bonds per atom

Hydrogen 1 needs to gain 1 electron so forms 1 covalent bond per atom

Ammonia NH_3

Oxygen 2,6 needs to gain 2 electrons so forms 2 covalent bonds per atom

Carbon 2,4 needs 4 electrons so forms 4 covalent bonds per atom and because it needs to gain more electrons it is the central atom of the molecule

Carbon dioxide CO_2

$O=C=C$

Each oxygen shares two pairs of electrons with the carbon and this means there are double covalent bonds between the carbon and each oxygen

NAILIT!

If you are doing the Foundation exam, concentrate on drawing dot-and-cross diagrams for compounds containing hydrogen. For example H_2O and NH_3.

If you are doing the Higher exam, you may be asked to draw dot-and-cross diagrams for compounds containing hydrogen, as well as for more complicated covalent bonding such as that shown in carbon dioxide where multiple bonds are involved (see Snap it! box).

Remember only the outer electrons are shown in dot-and-cross diagrams. The element which has to gain the most electrons is always the central atom in a molecule.

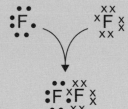

Some elements also exist as simple diatomic molecules. These elements are H_2, O_2, and N_2 and all the halogens (see Group 7) F_2, Cl_2, Br_2, I_2 and At_2. This is because they can form stable electron arrangements by bonding with each other.

For example, fluorine F_2.

CHECKIT!

1 Explain what is meant by a covalent bond.

2 What type of element has atoms that form covalent bonds?

3 Draw a molecule of water (H_2O) using a dot-and-cross diagram and using straight lines for the covalent bonds.

4 Draw dot-and-cross diagrams and line diagrams for

   a HF   b CF_4

5 CF_4 is a compound with a simple molecular structure. Predict some of its physical properties such as melting point, electrical conductivity, etc.

Diamond, graphite and graphene

All of the carbon atoms in diamond, graphite and graphene are linked together by strong covalent bonds.

Diamond and graphite exist as giant covalent structures.

In diamond, each carbon atom is covalently bonded to four other carbon atoms in a giant repeating pattern.

Each covalent bond is strong and difficult to break and this makes diamond **very hard**.

Because diamond has strong covalent bonds in a giant structure it has very high melting and boiling points.

There are no charged particles in diamond or any free electrons to carry an electric current so it does not conduct in the liquid or solid state.

In graphite the carbon atoms are in layers of hexagonal rings where each carbon atom is covalently bonded to three other carbons. This means that each carbon has a spare electron which is free to move through the layers so graphite conducts electricity both as a solid and as a liquid.

The covalent bonds in the layers are strong but between the layers there are only weak intermolecular forces. These are easy to break and so the layers can slide over each other easily, making graphite very soft and slippery. This means it can be used in industry as a lubricant.

Graphite also has high melting and boiling points because to melt graphite all of its strong covalent bonds have to be broken and this requires lots of energy.

Graphene has an identical structure to a single layer of graphite. This means that it is one atom thick.

It also means that graphene has identical properties to one layer of graphite, making it strong, flexible, transparent and a good electrical conductor. These properties give it lots of potential uses in, for example, electrical circuits, medical applications, displays and solar cells.

DOIT!

Draw a table with three columns: Physical property, Simple molecular and Giant covalent. Compare the physical properties of both structural types.

Physical property	Simple molecular	Giant covalent
Diamond		
Graphite		
Graphene		

SNAP**IT!**

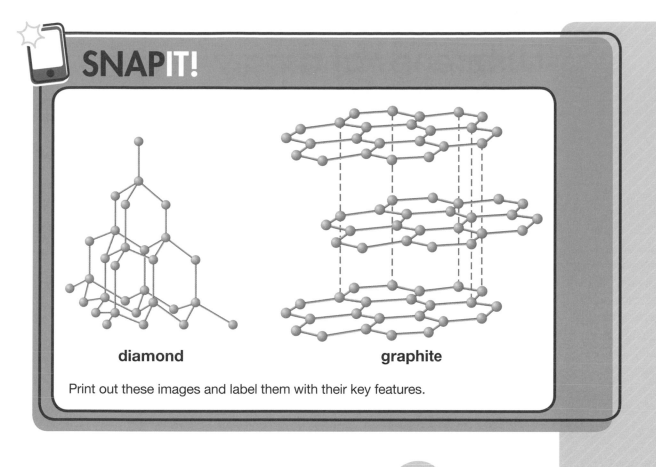

diamond   **graphite**

Print out these images and label them with their key features.

Diamonds

Graphite

NAIL**IT!**

You should be able to recognise the diagrams of diamond and graphite and understand that they are examples of giant covalent structures.

Describing or recognising the structures of diamond and graphite is not a high-level skill and would only get you a few marks in a question. The same applies to simply describing the properties. It is using the structure to explain the bulk properties that is a high-level skill.

✓ CHECK**IT!**

1 What type of structure is used to describe diamond and graphite?

2 Why are the melting points of diamond, graphite and graphene very high?

3 Explain why graphite is very soft.

4 Explain why diamond does not conduct electricity but graphite does.

5 Graphene is one layer of graphite. Describe this layer.

6 Silicon dioxide has a structure which is similar to diamond. Predict the properties of silicon dioxide.

Fullerenes and polymers

The smallest example of a spherical fullerene is Buckminsterfullerene which is a large but simple molecular substance with the formula C_{60}. These spheres contain mostly 6-carbon rings but there are also small numbers of 5- and 7-carbon rings.

These spherical fullerenes can be used to trap drugs and deliver them to parts of the body. They are also used in industry as lubricants and catalysts.

Carbon nanotubes are cylindrical fullerenes which have a high length to diameter ratio. Because of the strong covalent bonds between the carbons in the layers they have a high tensile strength. They are good electrical and heat conductors.

Polymers are very large molecules which are made up of long chains of carbon atoms joined by strong covalent bonds. Because the chains are very long the intermolecular forces between each chain are quite large and this means polymers are solids.

DOIT!

You may have used a key in biology to identify plants or animals. You can use a key to identify lots of different things. A key will usually ask questions based on easily identifiable features of something.

Draw out a key for carbon structures which can be used to identify them using the structures and properties. The key should include graphite, diamond, polymers, graphene and fullerenes (spherical and nanotubes).

SNAPIT!

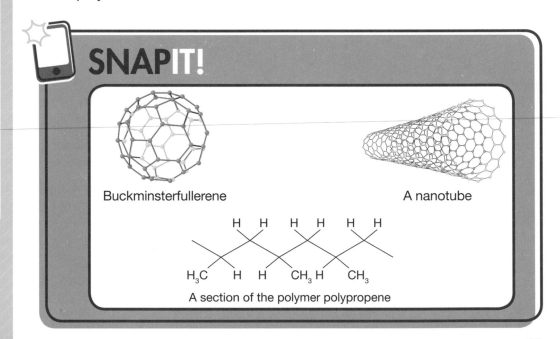

Buckminsterfullerene   A nanotube

A section of the polymer polypropene

CHECKIT!

1 a State how many carbon atoms there are in the smallest molecule of the Buckminsterfullerene series of structures.

   b This substance has a simple molecular structure. Predict some of its physical properties.

2 Describe the properties of nanotubes.

3 Explain why nanotubes are used to reinforce tennis rackets.

4 Give a use for Buckminsterfullerene.

Giant metallic structures and alloys

Metals form a giant lattice of positive metal ions in a sea of delocalised electrons.

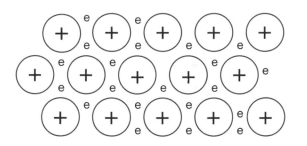

The electrostatic attraction between the positive ions and the delocalised or free electrons is called a metallic bond. These metallic bonds are usually strong and need a lot of energy to overcome them, which is why most metals have high melting and boiling points.

The metallic bond operates in both pure metals and in alloys.

The layers of metal ions can slide over each other without disrupting the structure. This means that metals can be bent, shaped (they are malleable) and drawn into wires (they are ductile).

The delocalised electrons can carry an electric current so metals are good electrical conductors both as solids and liquids.

Metals are also good heat or thermal conductors because the electrons can transfer the heat along the metal.

Alloys are mixtures of metals.

Introducing different-sized metal atoms into a metal lattice makes it harder for the layers to slide over each other so alloys are harder than pure metals.

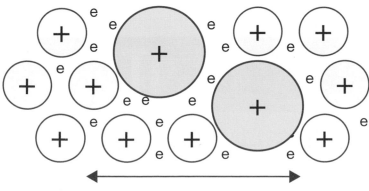

More difficult for metal ions to slide over each other

Car wheels are often made from alloys of magnesium or aluminium

DOIT!

Make a summary table for structure and fill it in. The headings are shown below:

Type of structure	Example	Particles present	Melting point and boiling point	Electrical conductivity	
				As solid	As liquid
Giant ionic					
Simple molecular					
Giant covalent					
Giant metallic					

NAILIT!

A common question is recognising the type of structure from the properties of a substance. These properties are usually melting and boiling points and electrical conductivity in the solid and liquid states.

SNAPIT!

Property	For metals	Explanation
Melting and boiling point	Usually high	Strong metallic bonds in a giant lattice
Electrical conductivity as solid	Good	Delocalised electrons carry the current
Electrical conductivity as liquid	Good	Delocalised electrons carry current
Thermal (heat) conductivity	Good	Delocalised electrons transfer the heat energy
Malleability and ductility	Malleable (can be shaped) and ductile (can be drawn into wires)	Layers of metal ions can slide over each other without changing the structure

CHECKIT!

1 Describe a giant metallic lattice.

2 Explain why metals are good electrical conductors in both the solid and liquid form.

3 Explain the term alloy.

4 Explain why alloys are often harder than pure metals.

Nanoparticles

A nanometre (nm) is 1 billionth of a metre or 1×10^{-9} m.

Nanoparticles are 1 nm to 100 nm in size.

To get some idea of the size of nanoparticles, coarse particles or dust have diameters 1×10^{-5} m (10000 nm) and 2.5×10^{-5} m (25000 nm). This means that a small particle of dust has a diameter 100 times the diameter of the largest nanoparticle.

Fine particles have diameters between 100 nm and 2500 nm.

Nanoparticles have very high surface area to volume ratios.

Their large surface area means that they will react very quickly. It also means that they make very good catalysts.

Nanoparticles also have very different properties to the same substance in larger sized particles. For example, titanium dioxide is a dense white solid used in house paint because of its white colour. Titanium dioxide nanoparticles are so small they are transparent and do not reflect visible light. They are used in sunscreens.

Uses of nanoparticles include controlled delivery of drugs; cosmetics and sunscreen and antibacterial agents in clothing.

DO IT!

Research three uses of nanoparticles, especially noting the differences between the properties of the nanoparticles and the normal bulk properties.

WORKIT!

Suppose you have two cubes. The first cube has sides 10 nm. The second smaller cube has sides equal to 1 nm. Work out the surface area to volume ratio of two cubes and explain what this tells us about the relationship between the size of particles and their surface area to volume ratio.

Cube 1 — its sides have an area of 100 nm^2. There are 6 sides so its total surface area is 600 nm^2.
Its volume = $10 \times 10 \times 10$ nm^3 = 1000 nm^3.
The surface area/volume ratio = 600/1000 = 0.6.

Cube 2 — the second smaller cube has sides equal to 1 nm. Its total surface area = 6 nm^2
Its volume is $1 \times 1 \times 1$ = 1 nm^3.
The surface area/volume ratio = 6/1 = 6. This is 10 times the surface area to volume ratio when the sides of the cube were 10 times bigger.

This shows that smaller particles have a greater surface area to volume ratio.

As the side of the cube decreases by a factor of 10 the surface area to volume ratio increases by a factor of 10.

MATHS SKILLS

You have to be aware of the high surface area to volume ratio of nanoparticles. To do this you must know how to work out areas and volumes.

If you have a cube with sides of length L, the area of each of its faces is L^2 and its volume is L^3.

CHECK IT!

1 Describe the range of sizes for nanoparticles.

2 List three uses for nanoparticles.

3 A cube has sides equal to 4 nm. What is its surface to volume ratio? What does this change to if the cubes are made smaller with 2 nm sides?

1 Explain how a lithium ion is formed from a lithium atom. [Atomic number of lithium = 3]

2 a State the size range for nanoparticles.

b List three uses for nanoparticles.

3 When magnesium (atomic number 12) reacts with chlorine (atomic number 17) the ionic compound magnesium chloride is formed.

a Showing the outer electrons only, draw dot-and-cross diagrams of magnesium and chloride ions.

b State the formula of magnesium chloride.

c Explain why magnesium chloride has a high melting point.

d Why is solid magnesium chloride a poor electrical conductor?

4 Carbon (atomic number 6) and hydrogen (atomic number 1) combine to form methane (CH_4).

a Draw a dot-and-cross diagram to show the bonding in methane.

b Explain why methane is a gas at room temperature even though the bonds between the carbon and hydrogen are strong.

5 These diagrams show the structures of diamond and graphite.

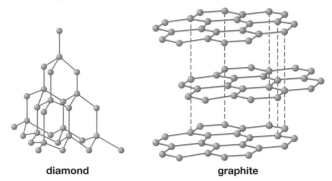

diamond   graphite

Answer the following questions about diamond and graphite:

a What type of structure do they both have?

b Why do they both have high melting points?

c Why is graphite slippery?

d Explain why solid graphite is a good electrical conductor but solid diamond is a poor electrical conductor.

6 The table below shows the properties of four substances. The letters are not their symbols.

Using the data in the table name the type of structure present in A to D.

Substance	Melting point /°C	Electrical conductivity	
		As solid	As liquid
A	−55	Poor	Poor
B	2015	Poor	Good
C	1897	Good	Good
D	2567	Poor	Poor

7 Using your knowledge of types of structure, explain the following observations:

a Methane has a lower melting and boiling point than potassium chloride.

H b Magnesium oxide has a higher melting point than potassium chloride.

c Brass (an alloy of copper and zinc) is harder than pure copper.

d Sodium conducts electricity in both the solid and liquid state but sodium chloride only conducts electricity in the liquid state or in a solution.

Quantitative chemistry

Conservation of mass and balancing equations

In chemical equations the reactants are written on the left-hand side and the products on the right-hand side.

Chemical formulae are used to represent the substances in the reactants and the products.

The law of conservation of mass states that no atoms are lost or gained during a chemical reaction.

Therefore, when you represent chemical reactions by symbol equations there must be the same number of each type of atom on both sides of the equation. This means that the equation must be **balanced**.

One consequence of this is that the mass of the reactants is equal to the mass of the products.

If the chemical container is open and gas is consumed or produced in a reaction, the mass appears to go up or down respectively but in terms of the atoms taking part it really has not changed.

Example:

$CaCO_3(s) + 2HNO_3(aq) \rightarrow Ca(NO_3)_2(aq) + H_2O(l) + CO_2(g)$

The mass appears to diminish because the CO_2 gas comes off from the reaction.

SNAPIT!

The diagram below shows apparatus that could be used to prove that the law of conservation of mass is true. There is a bung in the top of the flask so that no substances can enter or leave.

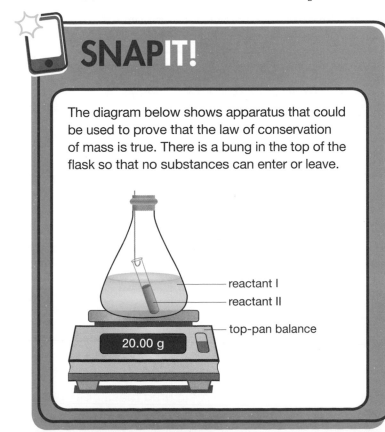

reactant I
reactant II

top-pan balance

20.00 g

DOIT!

Take the following equation and explain to a friend/revision partner why the equation has to be balanced and show them how it should be balanced. If working alone, write down your notes and check against your textbook.

$Fe_2O_3 + Al \rightarrow Fe + Al_2O_3$

NAILIT!

Balancing equations and being able to interpret them is one of the vital skills you need to get a good grade in chemistry. One thing that you may have to do is to balance equations that have in them negative acid groups such as nitrate (NO_3^-) and sulfate (SO_4^{2-}) or positive ammonium (NH_4^+). You must remember that if you have more than one of these groups in a compound then brackets are placed round the group.

For example, calcium nitrate contains the Ca^{2+} ion and the NO_3^- ion. Calcium nitrate's formula is therefore $Ca(NO_3)_2$.

When balancing equations like this treat the nitrate group like a single atom except when there is more than one, then they need brackets round them.

WORKIT!

The reaction between ammonia and copper(II) oxide gives nitrogen, copper and water as products. Write a balanced symbol equation for this reaction.

The unbalanced symbol equation may be written as
$NH_3 + CuO \rightarrow N_2 + Cu + H_2O$.

A good habit to get into is to follow each formula by the state symbol. NH_3 and N_2 are both gases so are followed by (g); CuO and Cu are solids, followed by (s) and H_2O is a liquid (l).

The number of nitrogen and hydrogen atoms are unequal and so the equation needs balancing. To balance it you can balance the nitrogen atoms first to give 2 on each side. Then the hydrogen atoms – 6 on each side – followed by the oxygen atoms – 3 on each side and the copper atoms – 3 on each side.

The final equation is
$2NH_3(g) + 3CuO(s) \rightarrow N_2(g) + 3Cu(s) + 3H_2O(l)$

MATHS SKILLS

A balanced symbol equation has the same number of each type of atom in the reactants as there are in the products.

When you balance a chemical equation the first thing you must remember is **not to change the formulae!**

You can only write numbers in normal-sized writing **before the formulae.**

Stage 1 - work out which atoms are not equal in number on both sides of the equation. Balance these by putting numbers in the correct places.

Stage 2 - see which atoms are now wrong in number and balance them and continue until the atoms are equal in number on both sides.

CHECKIT!

1 When magnesium is heated with aluminium oxide, magnesium oxide and aluminium are formed.

   a In this reaction identify which are the reactants and which are the products.

   b If 72 g of magnesium completely reacts with 103 g of aluminium oxide, what is the total mass of aluminium and magnesium oxide formed?

2 a Write the word equation for the following **unbalanced** chemical reaction:

   $HCl + CaCO_3 \rightarrow CaCl_2 + H_2O + CO_2$

   b Write out the balanced equation.

   c Add state symbols to the equation.

   d If this reaction was carried out in an open container the measured mass goes down. Explain why.

Relative formula masses

The relative atomic mass (symbol = A_r) of an element is the weighted average mass of its naturally occurring isotopes.

You calculate the relative formula mass (symbol = M_r) of a compound by adding up all the relative atomic masses of all the atoms present in the formula of the compound.

The elements hydrogen, oxygen, nitrogen, chlorine, bromine, iodine and fluorine exist as diatomic molecules. This means that in equations their relative formula masses are twice their relative atomic masses.

All other elements are represented by just their symbols in an equation. So their relative formula mass is equal to their relative atomic mass.

Using the law of conservation of mass we can say that in a chemical reaction the sum of the relative formula masses of the reactants is equal to the sum of the relative formula masses of the products.

SNAP IT!

Reactants	⟶	Products
sum of all the M_rs of reactants	=	sum of all the M_rs of products

For example when calcium carbonate (formula mass = 100) is heated it decomposes to calcium oxide (formula mass = 56) and carbon dioxide. This means that the formula mass of carbon dioxide equals 44.

MATHS SKILLS

In a chemical formula the number to the right-hand side of an atom is the number of that type of atom in the compound. For example, in carbon dioxide CO_2 there is 1 carbon atom and 2 oxygen atoms.

Also when there is a group of atoms such as hydroxide and there are brackets around the group in the formula then you multiply whatever is inside by the number outside it.

For example, calcium hydroxide $Ca(OH)_2$ has 1 calcium, 2 oxygens and 2 hydrogens.

NAIL IT!

The elements hydrogen, oxygen, nitrogen, chlorine, bromine, iodine and fluorine exist as diatomic molecules. Which means that in equations hydrogen is written as H_2, oxygen as O_2, etc. They can be remembered as HONClBrIF! Say it as a way of remembering it.

DO IT!

If you do not like using HONClBrIF to remember the elements that exist as molecules of 2 atoms (diatomic molecules) then make up your own mnemonic as an aid to memory.

WORKIT!

a Find the relative formula mass of aluminium oxide Al_2O_3.

There are 2 aluminium atoms and 3 oxygen atoms. The relative formula mass = $(2 \times 27) + (3 \times 16) = 102$.

b Find the relative formula mass of calcium nitrate $Ca(NO_3)_2$.

In this formula there is 1 calcium atom and 2 times whatever is inside the brackets. This means there are 2 nitrogen atoms and 6 oxygen atoms so the relative formula mass = $(1 \times 40) + (2 \times 14) + (6 \times 16) = 164$.

NAILIT!

You do not have to remember relative atomic masses of elements. You will be given a periodic table showing them. Make sure you use the right number. The relative atomic mass is the top one.

CHECKIT!

1 State the relative formula masses of the following compounds:

  a CaO

  b $MgCl_2$

  c KNO_3

  d $Al_2(SO_4)_3$

2 a The balanced equation below represents the reaction taking place when silver carbonate is heated:

$$2Ag_2CO_3(s) \rightarrow 4Ag(s) + 2CO_2(g) + O_2(g)$$

  b If 552 g of silver carbonate is heated, 88 g of carbon dioxide and 32 g of oxygen are formed. Calculate the mass of silver produced by the reaction.

  c What law are you applying in your calculation?

The mole

The **amount** of a chemical substance is measured in **moles** (symbol = mol).

The number of atoms, ions or molecules in a mole is equal to **Avogadro's constant** (N_A). The value of this is 6.02×10^{23}.

This number applies to all particles. For example, there are the same number of CH_4 molecules in 1 mole of methane as there are sodium atoms in 1 mole of the element sodium.

The mass of 1 mole of any substance is its relative formula mass expressed in grams.

The number of moles of a substance is equal to its mass in grams divided by its relative formula mass.

SNAPIT!

The equations for moles and number of particles

$$m = n \times M_r \qquad \text{no. of particles} = n \times N_A$$

$$n = m \div M_r$$

$$M_r = m \div n$$

There are two ways of learning the equations – either using the triangles or learning one equation and then rearranging it.

Consider the left-hand triangle. You have to use the operations shown in the triangle. This means that the top quantity m = the product of the two bottom quantities $n \times M_r$. The bottom left-hand quantity n = the top quantity m divided by the bottom left, etc.

DOIT!

Switch on your calculator and put it into Science mode. Use 3 as you will be asked to express to 3 significant figures.

Test out your understanding of putting numbers in standard form by writing 0.000065 in standard form.

Type 0.000065 into your calculator then press =. You should get 6.5×10^{-5}. Were you correct?

On some calculators you may need to press S↔D before getting your answer.

Test yourself on some other numbers.

MATHS SKILLS

Use the calculation triangles to construct formulae for your calculations.

$$n = m/M_r \qquad m = n \times M_r \qquad \text{no. of particles} = n \times N_A$$

When using Avogadro's constant you need to know how to use your calculator.

Type in 6.02 then press the [x10ˣ] button and then 23. You will get 6.02×10^{23}

NAILIT!

Know the relationships:
$n = m/M_r$
$m = n \times M_r$
number of particles $= n \times N_A$.

You need to be able to write down large numbers like Avogadro's constant in **standard form**.

The same applies to very small numbers. For example 0.00041 is expressed as 4.1×10^{-4}.

WORKIT!

How many moles of carbon dioxide (CO_2) are there in 2.2 g of the substance?

The relative formula mass of carbon dioxide = 44

This means that (see left-hand triangle on page 49)
$n = m/M_r = 2.2/44 = 0.05 \, mol$

How many carbon dioxide particles are there in the same mass of the substance?

See right-hand triangle on page 49

Number of CO_2 particles = number of moles × Avogadro's constant $= 0.05 \times 6.02 \times 10^{23} = 3.1 \times 10^{22}$

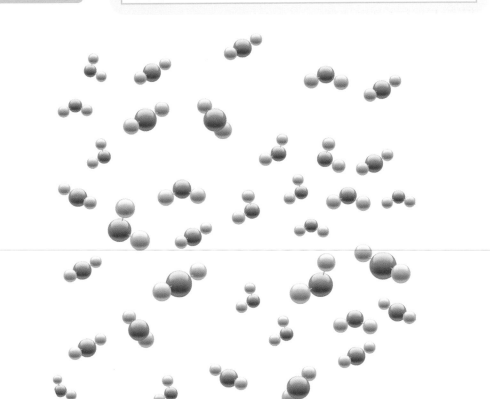

CHECKIT!

H 1 How many particles of a substance are found in 1 mole of the substance?

H 2 If you were given the formula of a compound how would you find the mass of 1 mole of the substance?

H 3 a   What is the relative formula mass of sulfur dioxide (SO_2)?

   b   How many moles are there in 1.6 g of sulfur dioxide?

   c   How many sulfur dioxide molecules are there in 1.6 g of the substance?

Reacting masses and using moles to balance equations

You can tell how many moles of a substance react or are produced in a reaction from the numbers in front of the formulae in the balanced symbol equation.

When you are given the masses that react you can convert these into moles and then use the equation to find the masses of the substances produced by using the ratios in the equation.

The same applies to finding the mass of reactants needed to produce a certain mass of products.

If you know the masses of reactants and products in a reaction you can convert these to moles and use the results to balance the equation.

DOIT!

Have a go at calculating one of the Work It! questions yourself. Record yourself explaining how you would work it out and compare to the solution given.

SNAPIT!

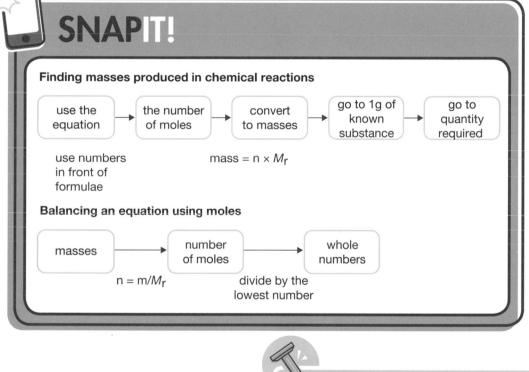

Finding masses produced in chemical reactions

use the equation → the number of moles → convert to masses → go to 1g of known substance → go to quantity required

use numbers in front of formulae

$mass = n \times M_r$

Balancing an equation using moles

masses → number of moles → whole numbers

$n = m/M_r$

divide by the lowest number

MATHS SKILLS

1 To work out the **mass of a substance produced** see the first worked example.

2 To see how to **balance an equation** see the second worked example.

NAILIT!

When you answer this type of numerical question and you are using a calculator, it is sometimes tempting to just write down the answer. This is not a good idea because if you make a mistake using the calculator and give the wrong answer without any working you will not get any marks. Always show your working and you will get credit for the correct approach.

You should also make sure your calculator will give an answer that enables you to quote your answer to 3 significant figures and will give the answer in standard form. For example:

$45 \div 4789$

If you use Scientific mode there is no problem. You can ask for 3 figures and you get 9.40×10^{-3}. So make sure you have it in this setup.

WORKIT!

1   Mass of a substance produced.

Calculate the mass of silver formed when 23.2 g of silver oxide is heated.

NOTE: Formula masses $Ag_2O = 232$; $Ag = 108$.

1.	Write the balanced equation.	$2Ag_2O$	$\longrightarrow$ 4Ag $\quad + \quad O_2$	
2.	Use the equation to find the number of moles of reactants and products.	2 moles of silver oxide	give	4 moles of silver	You are not asked about the oxygen.
3.	Convert these moles to masses.	$2 \times 232g =$ 464g	give	$4 \times 108g$ $= 432g$	Remember a mole is the M_r in g.
4.	IMPORTANT Go down to 1g of reactant to make calculation easier.	1g	give	432/464 g	Divide both sides by 464.
5.	Go to the desired quantity.	23.2g	give	$\dfrac{23.2 \times 432}{464}$ g $= 21.6g$	This will give 23.2 times whatever 1g gives.

2   Balancing equations using moles.

In the reaction between aluminium oxide and magnesium, 10.2 g of aluminium oxide reacts with 7.2 g of magnesium to give 5.4 g of aluminium and 12.0 g of magnesium oxide. Use these results to find the balanced equation for the reaction. [$M_r\,Al_2O_3 = 102$; $Mg = 24$; $Al = 27$ and $MgO = 40$]

Write word equation	Aluminium oxide +	magnesium $\longrightarrow$	aluminium	+ magnesium oxide
Write down the masses	10.2g	+ 7.2g	5.4g	12.0g
Convert these masses to moles	$\dfrac{10.2}{102} = 0.1$ mol	$\dfrac{7.2}{24} = 0.3$ mol	$\dfrac{5.4}{27} = 0.2$ mol	$\dfrac{12.0}{40} = 0.3$ mol
Divide by the lowest number which is 0.1 to give simplest whole-number ratio	$\dfrac{0.1}{0.1} = 1$	$\dfrac{0.3}{0.1} = 3$	$\dfrac{0.2}{0.1} = 2$	$\dfrac{0.3}{0.1} = 3$
Write out the balanced equation	Al_2O_3	+ 3Mg $\longrightarrow$	2Al	+ 3MgO

CHECKIT! ✓

H 1 Calculate the mass of silver that could be formed from 6.35 g of copper when copper is added to silver nitrate solution:

$$Cu(s) + 2AgNO_3(aq) \rightarrow Cu(NO_3)_2(aq) + 2Ag(s)$$

H 2 When 68 g of silver nitrate ($AgNO_3$) is heated it decomposes to form 43.2 g of silver (Ag), 18.4 g of nitrogen dioxide (NO_2) and 6.4 g of oxygen O_2.

Use this data to write a balanced equation for the reaction. You must show all your working.

[Relative formula masses: $AgNO_3 = 170$; $Ag = 108$; $NO_2 = 46$; $O_2 = 32$]

Limiting reactant

If there are two reactants in a chemical reaction and there is an excess (more than needed) of one of them then the other reactant is called the limiting reactant.

The amounts of the products formed depend on the amount of the limiting reactant.

The limiting reactant is identified by comparing the number of moles of each reactant with those required by the balanced symbol equation.

At the end of the reaction the limiting reactant is completely used up and the reactant in excess remains along with the products.

SNAPIT!

Identifying the limiting reactant in a reaction

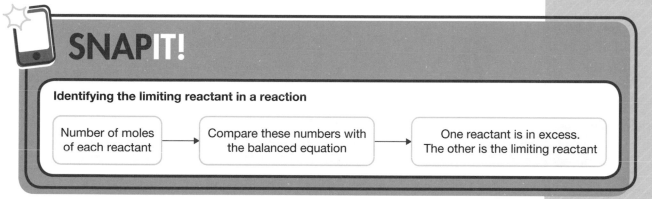

| Number of moles of each reactant | → | Compare these numbers with the balanced equation | → | One reactant is in excess. The other is the limiting reactant |

WORKIT!

The equation for the reaction between zinc and hydrochloric acid is
$Zn(s) + 2HCl(aq) \rightarrow ZnCl_2(aq) + H_2(g)$

If 0.10 mol of zinc is added to 0.3 mol of hydrochloric acid what is the limiting reactant?

From the equation 2 mol of hydrochloric acid react with 1 mol of zinc, and this means that 0.1 mol of zinc needs exactly 0.2 mol of hydrochloric acid to react.

This means that 0.3 mol of hydrochloric acid is in excess and the limiting reactant is zinc. Because there is 0.1 mol of zinc this means that we will get 0.1 mol of hydrogen and 0.1 mol of zinc chloride.

MATHS SKILLS

To identify the limiting reactant, you work out the number of moles of each reactant and then compare with what is needed using the chemical equation for the reaction.

CHECKIT!

H 1 Magnesium and sulfuric acid react as follows:

$Mg(s) + H_2SO_4(aq) \rightarrow MgSO_4(aq) + H_2(g)$

1.00 mol of magnesium is added to 0.90 mol of sulfuric acid.

a   Identify the limiting reactant.

b   Explain your answer.

c   How many moles of hydrogen are produced in the reaction?

Concentrations in solutions

DOIT!

Get all the 'calculation triangles' together and put them onto a sheet which you can put on your wall. Every so often when you take a break, look at the sheet. Gradually the information will sink in!

In a **solution** the substance being dissolved is called the **solute** and the liquid in which it dissolves is called the **solvent**.

A solution in water (a solvent) is described as an **aqueous** solution.

The greater the number of moles of solute dissolved in a solution the more concentrated it is. If the amount of water is increased then the solution becomes more dilute.

To compare the concentrations of solutions we use the amount of solute dissolved in $1\,dm^3$ ($1000\,cm^3$).

The concentration of a solution can be expressed in two different ways:

1   In grams of solute per dm^3 of solution. Units can be written as g per dm^3 or g/dm^3.

2   In moles of solute per dm^3 of solution. Units can be written as mol per dm^3 or mol/dm^3.

SNAPIT!

The equation triangles are shown below

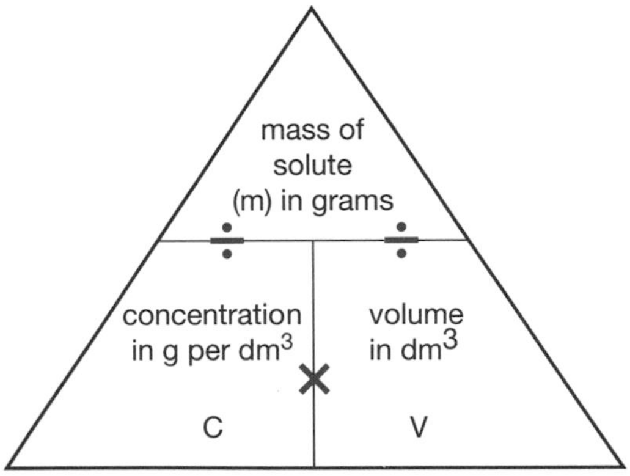

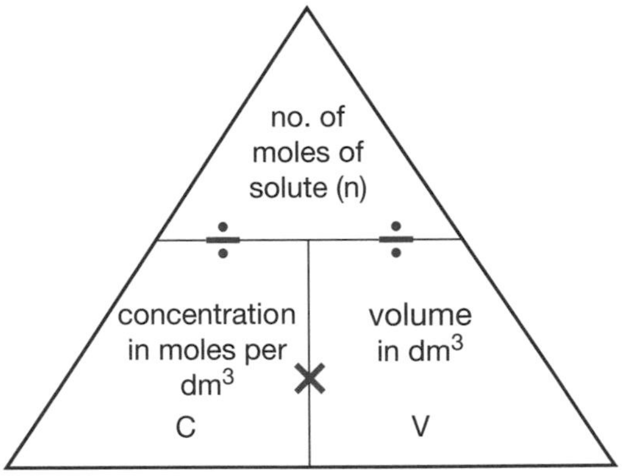

NAILIT!

Always convert the volume of a solution to dm^3 if it is given as cm^3.
$1\,cm^3 = 1/1000\,dm^3$
$= 1 \times 10^{-3}\,dm^3$
$= 0.001\,dm^3$.

MATHS SKILLS

Make sure you can use the formulae from these triangles:
$C = m/V$
$m = C \times V$
$V = m/C$
$C = n/V$
$n = C \times V$
$V = n/C$

WORKIT!

A solution contains 0.400 g of sodium hydroxide (M_r of NaOH = 40) in 100 cm^3 of solution. What is the concentration of the solution in both g per dm^3 and moles per dm^3?

Use the equations $C = m/V$; $n = m/M_r$ and then $C = n/V$.

$C = m/V$; $V = 100/1000\,dm^3 = 0.100\,dm^3$.
The concentration $= 0.400/0.1\,g$
per dm^3 $= 4.00\,g$ per dm^3.

Use $n = m/M_r = 0.4/40\,mol = 0.01\,mol$.
The concentration $= 0.0100/0.100\,mol/dm^3 = 0.100\,mol/dm^3$.

NAILIT!

If memorising the equation triangles is not your way of learning then think of how we express concentration. The units of concentration are mol per dm^3. Which is saying $C = n/V$. From there you can rearrange the equation to find V and n. The same applies to the formula $C = m/V$.

NAILIT!

As with all the other topics where calculations are made, make sure you show your working. Also when the question is longer highlight the part of the question that tells you what you have to find. See Check it! box below as an example.

✓ CHECKIT!

1 A solution of hydrochloric acid contains 0.1 mol in 500 cm^3.

What is its concentration?

a in mol/dm^3

b in g/dm^3

2 A solution has a concentration of 0.2 mol/dm^3. Calculate many moles of solute there are in 250 cm^3 of the solution.

Moles in solution

H

NAILIT!

Do not forget to convert the volumes of the solution from cm^3 to dm^3. If you are asked to find the volume of a solution do not forget that your answer will give you the number of dm^3 required. **Do not give it as cm^3 unless the question asks for the volume in cm^3 and then you must multiply your answer by 1000.**

Titrations are used to find the volumes of acids and alkalis that react together in a neutralisation reaction.

The reaction is complete when an indicator changes colour.

If you know the concentration of one of the two solutions and the volumes of both then you can find the concentration of the other solution.

The solution you know the concentration of is called the **standard solution.**

SNAP**IT!**

Using moles to find the concentration of a solution

In the descriptions below **standard solution** refers to the solution for which you have all the information – the concentration and volume. The **unknown solution** is the one you have to find. Firstly write the balanced symbol equation so that you know the ratio of moles of acid to alkali and vice versa.

| Use $n = C \times V$ to find number of moles of **standard** solution | → | Use equation to find number of moles of **unknown** solution | → | Use $C = n/V$ to find concentration of **unknown** solution |

Using moles to find the volume of a solution

| Use $n = C \times V$ to find number of moles of **standard** solution | → | Use equation to find number of moles of **unknown** solution | → | Use $V = n/C$ to find concentration of **unknown** solution |

MATHS SKILLS

Rearrange the equation for concentration with the three different quantities. See the Snap it! box for the procedure.

Practical Skills

One of the compulsory practicals for the Higher Tier is the titration of a strong acid against a strong alkali. You will also have to determine the concentration of one of these solutions using the known concentration of the other. This practical technique is covered on page 78.

WORKIT!

The concentration of a solution of sodium hydroxide is 1 mole per dm^3. In a titration experiment $20\,cm^3$ of this sodium hydroxide solution needed $40\,cm^3$ of sulfuric acid for complete reaction. What is the concentration of the sulfuric acid?

Procedure

1. Write the equation.
   $2NaOH(aq) + H_2SO_4(aq) \rightarrow Na_2SO_4(aq) + 2H_2O(l)$

2. Write down the information.

   $V = 20\,cm^3$
   $= 20/1000\,dm^3 =$
   $0.02\,dm^3$
   Concentration =
   $1\,mol$ per dm^3

   $V = 40\,cm^3 =$
   $40/1000\,dm^3$
   $= 0.04\,dm^3$
   Concentration
   unknown

3. Find the number of moles of NaOH.

   $n = C \times V = 1 \times 0.02$
   $= 0.02\,mol$

4. Use the equation to find the number of moles of the H_2SO_4 by using the molar ratios of acid to alkali.

   In the equation there are 2 of these to 1 of the H_2SO_4

   Number of mol
   $= \frac{1}{2} \times 0.02 =$
   $0.01\,mol$

5. Now use $C = n/V$ to find the concentration of the H_2SO_4.

   $C = n/V =$
   $0.01/0.04$
   $= 0.25\,mol/dm3$

CHECKIT!

H 1 Potassium hydroxide (KOH) and nitric acid (HNO_3) react with each other as follows:

$HNO_3(aq) + KOH(aq) \rightarrow KNO_3(aq) + H_2O(l)$

A solution of potassium hydroxide has a concentration of $0.200\,mol$ per dm^3 (or mol/dm^3). In an experiment to find the concentration of a solution of nitric acid, $20.0\,cm^3$ of the potassium hydroxide solution needed $30.0\,cm^3$ of the acid for complete reaction. Find the concentration of the nitric acid.

H 2 When hydrochloric acid is added to a solution of sodium carbonate the following reaction takes place:

$2HCl(aq) + Na_2CO_3(aq) \rightarrow 2NaCl(aq) + H_2O(l) + CO_2(g)$

In an experiment the concentration of the sodium carbonate was $0.500\,mol/dm^3$ and the concentration of the hydrochloric acid was $1.00\,mol/dm^3$. What volume of the acid is required to react completely with $20\,cm^3$ of the sodium carbonate solution?

H

Moles and gas volumes

At the same temperature and pressure equal volumes of different gases contain the same number of molecules.

This means that under the same conditions equal volumes of gases have the same number of moles present.

At room temperature (20 °C) and 1 atmosphere pressure (RTP), 1 mole of any gas occupies a volume of 24 dm^3 (24 000 cm^3).

SNAP IT!

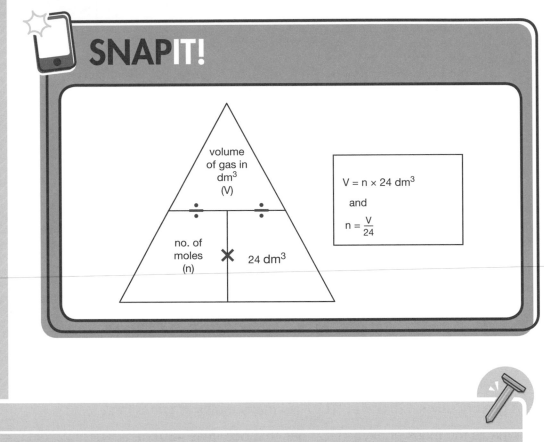

$$V = n \times 24 \text{ dm}^3$$

and

$$n = \frac{V}{24}$$

NAIL IT!

Remember all the rules we have about amounts of substances also apply to gases. This means that if we know the mass of a gas we can find its volume and if we know its volume we can find its mass.

(RTP = Room temperature and 1 atmosphere pressure)

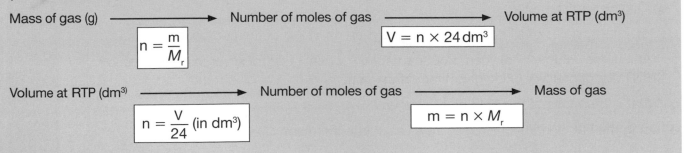

Mass of gas (g) ⟶ Number of moles of gas ⟶ Volume at RTP (dm^3)

$$n = \frac{m}{M_r}$$

$$V = n \times 24 \text{ dm}^3$$

Volume at RTP (dm^3) ⟶ Number of moles of gas ⟶ Mass of gas

$$n = \frac{V}{24} \text{ (in dm}^3)$$

$$m = n \times M_r$$

WORKIT!

A gas occupies 800 cm³ at RTP. How many moles of gas is this?

Use the equation $n = \dfrac{V}{24}$. Remember 24 refers to dm³ and therefore we have to convert the 800 cm³ to dm³. This means that $V = \dfrac{800}{1000}$ dm³ = 0.8 dm³. Therefore the number of moles of gas $(n) = \dfrac{0.8}{24}$ mol = 0.0333 mol or 3.33×10^{-2} mol.

When hydrochloric acid is added to calcium carbonate it forms a solution of calcium chloride, water and carbon dioxide gas.

$$2HCl(aq) + CaCO_3(s) \rightarrow CaCl_2(aq) + H_2O(l) + CO_2(g)$$

a Calculate the volume of carbon dioxide gas given off at RTP when 0.2 mol of hydrochloric acid is added to an excess of calcium carbonate. (See 'Limiting reactant' on page 53).

From the equation 2 mol of hydrochloric acid give 1 mol of carbon dioxide gas. This means that 0.2 mol of HCl will give 0.1 mol of CO₂. This amount of gas at RTP occupies 0.10×24.0 dm³ = 2.40 dm³.

b What is the mass of this volume of carbon dioxide?

Number of moles of gas = 0.10 mol. The relative formula mass of carbon dioxide = 44.

The mass of CO₂ = $n \times M_r$ = 0.10×44 g = 4.40 g.

✓ CHECKIT!

H 1 Give the formula to find the number of moles of a gas and its volume at RTP.

H 2 A gas occupies 120 cm³ at RTP How many moles of gas is this?

H 3 What is the mass of oxygen (O_2) gas that occupies 480 cm³ at RTP? [M_r of O_2 = 32]

H 4 When silver oxide (Ag_2O (M_r = 232)) is heated the following decomposition reaction takes place:

$$2Ag_2O(s) \rightarrow 4Ag(s) + O_2(g)$$

a How many moles of silver oxide (Ag_2O) are there in 11.6 g of the compound?

b What is the volume of oxygen produced at RTP when 11.6 g of silver oxide is heated? [HINT: 2 mol of silver oxide give 1 mol of oxygen]

c What is the mass of this volume of oxygen?

d Using the law of conservation of mass calculate the mass of silver produced.

Percentage yield and atom economy

There are a few reasons why the percentage yield drops to below 100%:

- Some of the product may be lost in the purification or separation process.

- The products may react back to give the reactants in a reversible process.

- There may be other reactions taking place that do not give the desired product.

The atom economy is a measure of how much of the reactants end up as useful products.

Atom economy = $\dfrac{\text{mass of wanted product(s)}}{\text{total mass of product(s)}}$ x 100%

If there is only one product then the atom economy is 100%.

$$\text{Percentage yield} = \frac{\text{mass of product actually made}}{\text{maximum theoretical mass of product}} \times 100\%$$

The maximum theoretical yield is the maximum mass of product that could be made.

The percentage yield is an indication of the efficiency of a chemical process.

WORKIT!

In an experiment to demonstrate the thermite reaction, excess aluminium reacted with 8.00 g of iron(III) oxide to give iron and 5.10 g of aluminium oxide.

The equation for the reaction is: $2Al(s) + Fe_2O_3(s) \rightarrow 2Fe(s) + Al_2O_3(s)$

a i What is the limiting reactant in this reaction?

The iron(III) oxide is the limiting reactant because the aluminium is in excess.

 ii Calculate the maximum theoretical yield of iron in the reaction.

From the equation 1 mol of iron(III) oxide gives 2 mol of iron.

 iii If 4.2 g of iron was formed, what is the percentage yield?

Therefore 160 g of iron(III) oxide gives $2 \times 56 = 112$ g of iron.

This means that 1 g of iron(III) oxide gives $\frac{112}{160}$ g of iron.

Therefore 8 g of iron(III) oxide should give $8 \times \frac{112}{160}$ g of iron $= 5.6$ g

The percentage yield $= \dfrac{\text{mass of product actually made}}{\text{maximum theoretical mass of product}} \times 100\% = \dfrac{4.2}{5.5}$

$\times 100\% = 75\%$

b What is the atom economy for this reaction if the desired product was iron?
 [M_r Al = 27; Fe_2O_3 = 160; Fe = 56; Al_2O_3 = 102]

Atom economy of a reaction $= \dfrac{\text{relative formula mass of desired product in the reaction}}{\text{sum of relative formula masses of reactants}}$
$\times 100\%$

In this reaction the atom economy $= \dfrac{2 \times 56}{2 \times 27 + 160} \times 100\% = \dfrac{112}{214} \times 100\%$

$= 52.3\%$

CHECKIT!

1 If the actual yield of product in a reaction is 24.0 g and the theoretical yield is 96.0 g, what is the percentage yield? Explain why it is not 100%.

2 The equations below show two methods to prepare ammonia (NH_3). Calculate the atom economy for both methods:

a $NH_4Cl(s) + NaOH(aq) \rightarrow NH_3(g) + NaCl(s) + H_2O(l)$

b $N_2(g) + 3H_2(g) \rightleftharpoons 2NH_3(g)$

Quantitative chemistry

For this set of questions you will need your periodic table. You will also need to use the following information:

Avogadro's constant (N_A) = 6.02×10^{23}; 1 mole of any gas occupies $24\,dm^3$ at RTP.

You should also refer to the formulae sheets that you have been making as we have gone through this topic. Have your list of formulae ready.

1 a Express the following numbers in standard form:

    i 0.0833   **ii** 223 000   **iii** 856.1
    iv 0.0000453

b Write the following numbers in standard form and to three significant figures.

    i 4   **ii** 0.06572   **iii** 0.04550
    iv 0.0004389567900

2 a Balance the following equations:

    i $H_2(g) + Cl_2(g) \rightarrow HCl(g)$
    ii $Na(s) + Br_2(l) \rightarrow NaBr(s)$
    iii $K(s) + N_2(g) \rightarrow K_3N(s)$
    iv $Mg(s) + AgNO_3(aq) \rightarrow Mg(NO_3)_2(aq) + Ag(s)$
    v $Na(s) + O_2(g) \rightarrow Na_2O(s)$

b Explain why these equations have to be balanced.

3 Give the relative formula masses of the following compounds:

    a K_2O **b** $Ca(OH)_2$ **c** $Mg(NO_3)_2$ **d** $Al(OH)_2$
    e potassium sulfate   **f** copper(II) chloride
    g silicon dioxide

H 4 a Calculate the relative formula mass of CO_2.

b How many moles of carbon dioxide are present in 4.4 g of CO_2?

c How many CO_2 molecules are there in 4.4 g of CO_2?

d What is the volume of this mass of CO_2 at RTP?

5 The equations below show two ways by which carbon dioxide can be prepared:

    i $2HCl(aq) + CaCO_3(s) \rightarrow CaCl_2(aq) + H_2O(l) + CO_2(g)$
    ii $C(s) + O_2(g) \rightarrow CO_2(g)$

a Calculate the atom economies of both methods.

b Why is atom economy important to chemists?

H 6 The equation below represents the reaction between sodium hydroxide and hydrochloric acid:

$HCl(aq) + NaOH(aq) \rightarrow NaCl(aq) + H_2O(l)$

In one experiment on this reaction $20.0\,cm^3$ of hydrochloric acid reacted with $30.0\,cm^3$ of NaOH which had a concentration of 1 mole per dm^3. What is the concentration of the acid?

H 7 Magnesium and hydrochloric acid react as follows:

$Mg(s) + 2HCl(aq) \rightarrow MgCl_2(aq) + H_2(g)$

6 g of magnesium are added to $200\,cm^3$ of hydrochloric acid with a concentration of 1 mole per dm^3.

a How many moles of magnesium are there in 6 g of the metal?

b How many moles of hydrochloric acid were used?

c Explain which of the two reactants was the limiting reactant.

Chemical changes

Metal oxides and the reactivity series

Most metals react with oxygen to form metal oxides.

Gain of oxygen is oxidation and loss of oxygen is reduction.

The reactivity series (see Snap it! box on the next page) is an arrangement of the metals in order of their reactivity. The non-metals carbon and hydrogen are also included in the series.

When metals react they lose electrons to form positive ions. The more reactive the metal the more easily it loses electrons.

Metals that react with water at room temperature give metal hydroxides and hydrogen gas is produced. For example, when calcium is added to water, calcium hydroxide is formed along with hydrogen.

$$Ca(s) + 2H_2O(l) \rightarrow Ca(OH)_2(aq) + H_2(g)$$

When metals react with dilute acids metal salts are produced and hydrogen gas is given off. For example, when magnesium is added to hydrochloric acid the salt formed is magnesium chloride and hydrogen is also given off.

$$Mg(s) + 2HCl(aq) \rightarrow MgCl_2(aq) + H_2(g)$$

More reactive metals will displace less reactive metals from solutions of metal salts.

For example, magnesium is more reactive than iron. This means that if magnesium is added to a solution of iron(II) sulfate the iron is displaced to give a solution of magnesium sulfate and iron.

$$Mg(s) + FeSO_4(aq) \rightarrow Fe(s) + MgSO_4(aq)$$

NAILIT!

Metal salts are formed when the hydrogen in an acid molecule is replaced by a metal atom. For example, sulfuric acid, H_2SO_4, has two replaceable hydrogens and either one or both of them can be replaced. This means that salts of sulfuric acid include $NaHSO_4$, Na_2SO_4 and $MgSO_4$.

Crystals of iron(II) sulfate

SNAPIT!

The Reactivity Series

This diagram shows the relative reactivity of some common metals. The way to remember the order is simply to say **PoSLiCaMZIC**. The letter(s) representing the metals are not necessarily their usual symbols.

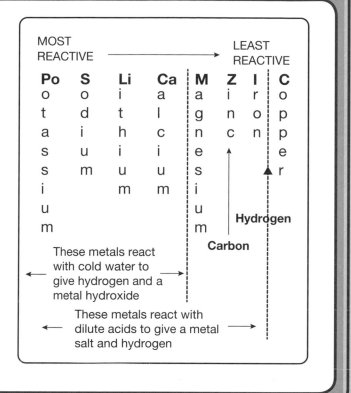

DOIT!

If you do not like the **phonetic** method of remembering the reactivity series (PoSLiCaMZIC), make up your own mnemonic.

CHECKIT!

1 What is formed when a metal atom reacts?

2 Which is more reactive, carbon or copper?

3 What is formed when lithium is added to water?

4 Write the word equation and balanced symbol equation for the reaction between hydrogen (H_2) and copper(II) oxide (CuO).

5 Write the word equation and balanced symbol equation for the reaction between magnesium (Mg) and copper sulfate ($CuSO_4$).

6 What is formed when dilute hydrochloric acid is added to:

a zinc metal

b copper metal?

Extraction of metals and reduction

DOIT!

One way of getting the reactivity series into perspective and to see why it is useful is to consider the uses of some of the metals. For example, copper is very unreactive which makes it useful for pipes because it will not react with the water. Consider some of the other metals on the list like magnesium and sodium.

STRETCHIT!

Aluminium is a reactive metal but can be used for cooking foil and aeroplane fuselages. Do some research to find out why.

Metals in the Earth's crust are found in rocks called ores. These ores contain enough of the metal to make it worthwhile to use them.

Unreactive metals such as gold and silver can be found uncombined with other elements and don't need chemical reactions to extract them.

If a metal is less reactive than carbon then it can be extracted by heating the metal oxide with carbon. The metal oxide is reduced by losing its oxygen.

SNAPIT!

Metals more reactive than carbon are usually extracted using electrolysis.

Potassium Sodium Lithium Calcium Magnesium Aluminium	→	These metals are extracted using **electrolysis** because they are more reactive than carbon and their oxides are **not reduced** by heating with carbon
Zinc Iron Copper	→	These 3 metals are less reactive than carbon and therefore their oxides can be reduced to the metal by heating with carbon
Silver Gold	→	Silver and gold are found uncombined (on their own) and do not have to be extracted

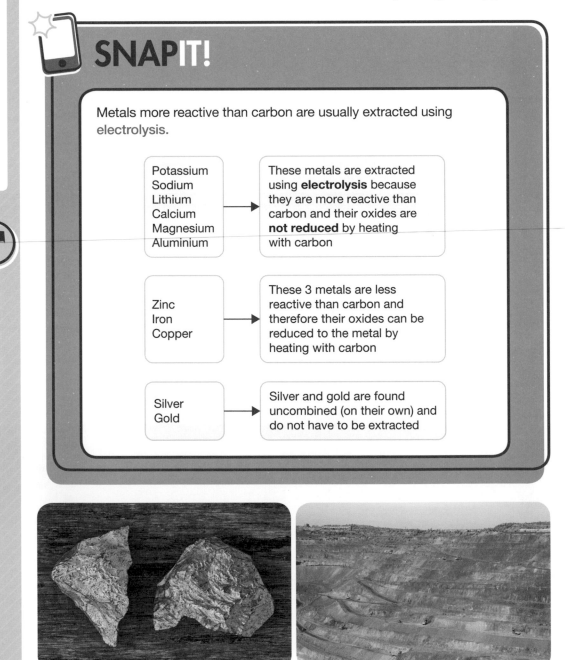

Gold and silver can be found uncombined with other elements, while other elements are found in ores

STRETCHIT!

In compounds, metals exist as positively charged ions. This means that when they react they lose electrons. Another definition of oxidation is that **Oxidation Is Loss** of electrons.

The more reactive the metal the more easily they lose electrons and the less easily they take them back.

In electrolysis, electrons are gained by metal ions to form metal atoms. This is another version of reduction. **Reduction Is Gain** of electrons.

We remember this definition of oxidation and reduction by using the mnemonic **OILRIG**.

Oxidation **I**s **L**oss of electrons and **R**eduction **I**s **G**ain of electrons.

Examples of oxidation involving metals losing electrons are:

$Fe \rightarrow Fe^{2+} + 2e^-$ and $Al \rightarrow Al^{3+} + 3e^-$

Examples of reduction are the reverse of those above and $Na^+ + e^- \rightarrow Na$.

NAILIT!

When metals oxides are reduced by carbon to the metal, carbon dioxide is usually formed.

For example copper(II) oxide and carbon form copper and carbon dioxide.

$2CuO(s) + C(s) \rightarrow 2Cu(s) + CO_2(g)$

When carbon is heated with the oxide of a more reactive metal, there is no reaction.

$Al_2O_3(s) + C(s) \rightarrow$ No reaction

CHECKIT!

1 What term is used for the following:

   a a metal gaining oxygen

   b a metal losing oxygen?

2 In the reaction shown below which metal is reduced and which metal is oxidised?

   $2Al(s) + Fe_2O_3(s) \rightarrow Al_2O_3(s) + 2Fe(s)$

3 a Lead is a more reactive metal than copper but less reactive than iron. How is lead extracted from its ore?

   b Barium is more reactive than calcium. How is barium extracted from its ore?

The reactions of acids

Acids are a group of substances with similar properties. This means that if you know the properties of one acid you can predict the properties of the others.

The reason all acids have similar properties is that they all contain hydrogen, and when in aqueous solution they all give hydrogen ions (H^+).

Hydrochloric acid (HCl) and sulfuric acid (H_2SO_4) react with metals to form salts and hydrogen.

Alkalis are soluble metal hydroxides; bases are insoluble metal oxides and hydroxides.

Examples of alkalis are sodium hydroxide (NaOH) and potassium hydroxide (KOH). Examples of insoluble bases are copper(II) oxide (CuO) and magnesium hydroxide ($Mg(OH)_2$).

Hydrochloric acid (HCl), nitric acid (HNO_3) and sulfuric acid (H_2SO_4) are all neutralised by alkalis and bases. Each neutralisation reaction forms a salt and water.

acid(aq) + alkali(aq) → salt(aq) + water(l)

acid(aq) + base(s) → salt(aq) + water(l)

All three of the acids above react with metal carbonates to give a salt, water and carbon dioxide gas. So the general equation is:

acid(aq) + carbonate(s) → salt(aq) + water(l) + carbon dioxide(g)

For example, calcium carbonate ($CaCO_3$) reacts with nitric acid (HNO_3) to produce the salt calcium nitrate ($Ca(NO_3)_2$), carbon dioxide (CO_2) and water (H_2O).

$$CaCO_3(s) + 2HNO_3(aq) \rightarrow Ca(NO_3)_2(aq) + CO_2(g) + H_2O(l)$$

H

The reaction between acids and metals is a redox reaction. The metal atoms lose electrons and are oxidised. The hydrogen (H^+) ions are reduced because they gain electrons.

e.g. $Mg(s) + 2H^+(aq) \rightarrow Mg^{2+}(aq) + H_2(g)$

Note that the negative ions associated with the acid do not appear in these equations because they do not take part in the reaction. They are spectator ions ('they just look on').

H

DO IT!

Make up an exam question about the reactions of acids for a friend or revision partner.

- Each part of the question should be worth a certain number of marks.
- Write a mark scheme for your question.

NAIL IT!

One common problem associated with the writing of symbol equations involving acids is the writing of the formulae of the salt because of the negative ions in acids like sulfuric acid and nitric acid.

- The key to this is realising that the acids contain replaceable hydrogen ions.
- For example sulfuric acid is H_2SO_4. It contains two replaceable hydrogen ions. Each of these has one positive charge. Therefore the sulfate ion must have two negative charges to balance out the positive charges from the hydrogen ions and its formula is SO_4^{2-}.
- If sulfuric acid forms a salt then we can use this information to write the correct formulae for the salt.
- For example, magnesium sulfate contains the magnesium Mg^{2+} ion (magnesium is in Group 2). Therefore the formula of magnesium sulfate is $MgSO_4$ as the +2 on the magnesium ion cancels out the -2 on the sulfate ion.
- Similarly the nitrate ion is NO_3^- [only one replaceable hydrogen ion in nitric acid (HNO_3)]. Therefore in magnesium nitrate we need two nitrate ions to cancel out the charges on the Mg^{2+}. Therefore magnesium nitrate is written $Mg(NO_3)_2$.

SNAPIT!

This table shows salts formed by metals with different acids. Remember to find the formula of the salt the charges on the ions have to be balanced (see Ions and ionic bonding on page 32.)

Metal in base; alkali etc.	Formula of metal ion	Acid	Ion from acid	Name of salt formed	Formula of salt
Magnesium	Mg^2	Nitric (NHO_3)	NO_3^-	magnesium nitrate	$Mg(NO_3)_2$
Copper	Cu^{2+}	Sufl (H_2SO_4)	SO_3^{2-}	copper(II) sulfate	C
Zinc	Zn^{2+}	Hydrochloric (HCl)	Cl^-	zinc chloride	$ZnCl_2$
Sodium	Na^+	Sulfuric (H_2SO_4)	SO_4^{2-}	sodium sulfate	Na_2SO_4

Practical Skills

There are two practicals associated with this topic. You can see them on pages 68 and 80. They are:

1 The preparation of a soluble salt by the reaction of a base with an acid.

2 The determination of the reacting volumes of a strong acid and an alkali by titration.

H 3 Using the results from the titration to find the concentration of either the acid or the alkali.

CHECKIT!

1 What is meant by the terms:

a alkali

b base?

2 Name the salts formed when sodium hydroxide reacts with:

a hydrochloric acid

b nitric acid

3 When copper carbonate is added to sulfuric acid, the mixture fizzes/effervesces. Explain this observation.

4 Nitric acid forms sodium nitrate ($NaNO_3$).

a Give the formula of the nitrate ion.

b Give the formula of magnesium nitrate.

5 Complete and balance the following symbol equations:

a $MgO(s) + H_2SO_4(aq) \rightarrow$

b $MgCO_3(s) + H_2SO_4(aq) \rightarrow$

c $Mg(OH)_2(s) + HCl \rightarrow$

H 6 a Give the ionic equation for the reaction between an acid and magnesium.

b Explain why this is a redox reaction.

Practical: The preparation of a soluble salt

A soluble salt is prepared by the neutralisation reaction between an insoluble base and an acid.

If a metal carbonate is used, then the only difference is that carbon dioxide is given off.

The metal ion needed to make the salt comes from the base and the negative ion in the salt comes from the acid.

The general equation for the reaction shows that the reaction gives a clear solution because there is only water and an aqueous solution of a salt formed.

Base(s) + acid(aq) → salt(aq) + water(l)

Therefore when the insoluble base stops dissolving we know all the acid has been used up.

Practical Skills

The skills/ techniques used are heating, filtration, evaporation and crystallisation.

Stage	What is done/action	Explanation of what is done
1	Heat up the acid.	Higher temperature means faster reaction.
2	Add the solid base.	Neutralisation reaction takes place between acid and base.
3	Stop adding the base when it stops dissolving or if a carbonate is added it stops fizzing.	When the base stops dissolving or the carbonate, stops fizzing it means that the acid has been used up.
4	Filter off the unreacted base or carbonate.	Remove unreacted insoluble base or carbonate, leaving just the aqueous solution of the salt.
5	Heat the salt solution on a steam bath.	Steam bath evaporates the water in the salt solution slowly so we get crystals and not powder.
6	Stop heating when solid salt starts to appear. You now have a saturated solution.	Saturated solution means that no more will dissolve in the solution at the higher temperature.
7	Leave the saturated solution to cool so that crystallisation can take place.	At a lower temperature the solid is less soluble so more crystals appear.

SNAPIT!

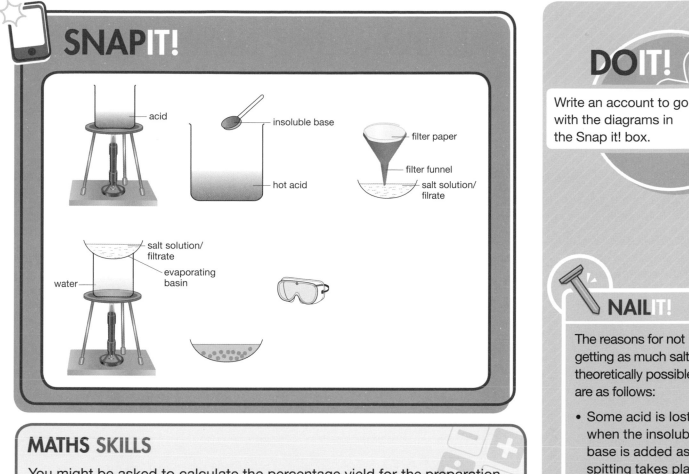

MATHS SKILLS

You might be asked to calculate the percentage yield for the preparation and account for any losses.

$$\text{Percentage yield} = \frac{\text{actual yield (what you get)}}{\text{theoretical yield (what you should get)}} \times 100\%$$

DOIT!

Write an account to go with the diagrams in the Snap it! box.

NAILIT!

The reasons for not getting as much salt as theoretically possible are as follows:

- Some acid is lost when the insoluble base is added as spitting takes place.
- Some salt solution is left in the filter paper.
- Not all of the salt solution crystallises.

CHECKIT!

1 What reactants could you use to prepare the following soluble salts?

  a Copper sulfate

  b Zinc nitrate

  c Magnesium chloride

2 a How is an unreacted base separated from the salt solution?

  b Why is this method of separation used?

3 In a salt preparation 4.5 g of the salt were prepared when 5 g was the theoretical yield. What was the percentage yield?

4 The preparation of magnesium nitrate can be made using two different reactions and the equations for these are shown below:

A $MgO(s) + 2HNO_3(aq) \rightarrow Mg(NO_3)_2(aq) + H_2O(l)$

B $MgCO_3(s) + 2HNO_3(aq) \rightarrow Mg(NO_3)_2(aq) + H_2O(l) + CO_2(g)$

  a Give the **two** ways we can tell that reaction B has been completed.

  b Assuming that magnesium nitrate is the desired product, calculate the atom economy for each method.

  [Relative formula masses MgO = 40; $MgCO_3$ = 84; HNO_3 = 63; $Mg(NO_3)_2$ = 148; H_2O = 18]

Oxidation and reduction in terms of electrons

In **displacement** reactions a reactive metal will displace a less reactive one from an aqueous solution of its salt.

As far as electrons are concerned, **O**xidation **I**s **L**oss of electrons and **R**eduction **I**s **G**ain of electrons (remembered as **OILRIG**).

Ionic equations for displacement reactions will only involve metal atoms and positive metal ions.

The ions of the less reactive metal gain electrons from the atoms of the more reactive metals.

This means that the less reactive metal ions are reduced and the more reactive metal atoms are oxidised.

The negative ions from the salt are **spectator ions** – they are not involved in the reaction and can be removed from the equation.

NAILIT!

When you write the ionic equations <u>leave out</u> the non-metal ions and just put in the metal atoms and ions present. For example, the reaction between zinc metal and an aqueous solution of copper(II) sulfate:

$Zn(s) + Cu^{2+}(aq) \rightarrow Zn^{2+}(aq) + Cu(s)$

Note that the zinc has lost electrons and become more positive by forming a Zn^{2+} ion (oxidised) and at the same time the copper(II) ion has lost its +2 charge by accepting electrons and forming a neutral copper atom (reduced).

SNAPIT!

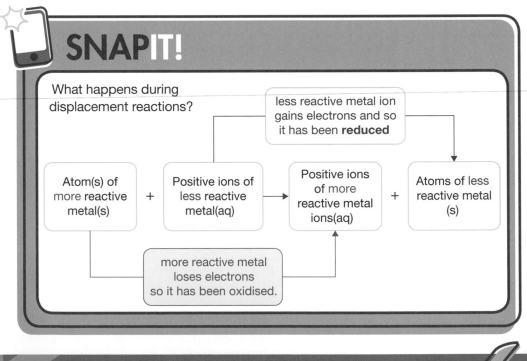

What happens during displacement reactions?

Atom(s) of more reactive metal(s) + Positive ions of less reactive metal(aq) → Positive ions of more reactive metal ions(aq) + Atoms of less reactive metal (s)

less reactive metal ion gains electrons and so it has been **reduced**

more reactive metal loses electrons so it has been oxidised.

CHECKIT!

H 1 What are the definitions of reduction and oxidation in terms of loss and gain of electrons?

H 2 a Write the following reaction as an ionic equation:

$Mg(s) + ZnCl_2(aq) \rightarrow MgCl_2(aq) + Zn(s)$

b What has been oxidised and what has been reduced in this reaction?

The pH scale and neutralisation

An aqueous solution is one formed when a substance dissolves in water. Its state symbol is (aq).

Acids are a group of substances with similar properties. This is because all aqueous solutions of acids produce H^+ ions.

In the same way aqueous solutions of alkalis all give OH^- ions.

In neutralisation reactions H^+ and OH^- ions react to give water. The ionic equation for this reaction is as follows:

$$H^+(aq) + OH^-(aq) \rightarrow H_2O(l)$$

The pH scale runs from 0 to 14 and is a measure of the acidity or alkalinity of a solution.

Acid solutions have a pH value less than 7. The lower the pH the more concentrated the H^+ ions are.

Neutral solutions such as pure water have a pH of 7.

SNAP IT!

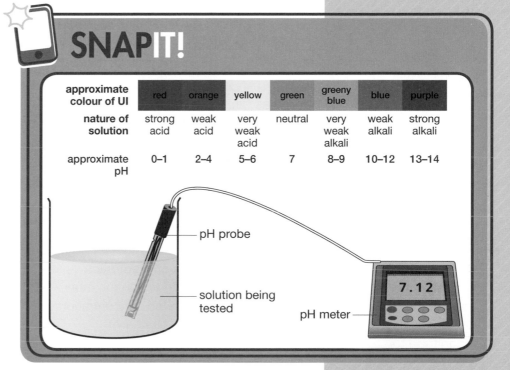

approximate colour of UI	red	orange	yellow	green	greeny blue	blue	purple
nature of solution	strong acid	weak acid	very weak acid	neutral	very weak alkali	weak alkali	strong alkali
approximate pH	0–1	2–4	5–6	7	8–9	10–12	13–14

pH probe
solution being tested
pH meter
7.12

Alkaline solutions have a pH value greater than 7 and the higher the pH the greater the concentration of OH^- ions.

The pH of a solution can be measured accurately using a pH probe or more approximately using the colour of universal indicator (UI) in the solution.

A pH probe measures the concentration of hydrogen ions accurately and the pH is read from a pH meter either digitally or on a scale.

CHECK IT!

1 What ions would you find in aqueous solutions of both hydrochloric acid and sulfuric acid?

2 What ions would you find in aqueous solutions of both sodium hydroxide and calcium hydroxide?

3 When hydrochloric acid is neutralised by an alkali what reaction takes place?

4 The pH of a solution of hydrochloric acid is 1. Ethanoic acid of the same concentration has a pH of 3. What do these observations tell you about the two acids?

Strong and weak acids

DO IT!

Use your calculator to express concentrations (mol per dm^3) in standard form. The numbers you need to look at are 1, 0.1, 0.001, etc.

Describe what happens to the order of magnitude n, in 1×10^{-n} as the concentration decreases 10-fold each time.

A concentrated solution of an acid has a greater amount (in moles) of the acid dissolved in the same volume of water than a dilute solution.

A strong acid like hydrochloric acid is completely ionised in aqueous solution.

A weak acid like ethanoic acid is only partially ionised in aqueous solution.

This means that 1000 molecules of the strong acid, hydrochloric acid, will all ionise to give 1000 H^+ ions in aqueous solution. On the other hand, only 4 out of 1000 molecules of the weak acid, ethanoic acid, will ionise in aqueous solution.

IMPORTANT – strong is not the same as concentrated and weak is not the same as dilute.

If the pH value of a solution decreases by 1 unit then the concentration of the H^+ ions increases by 10 times or 1 order of magnitude.

WORKIT!

As the pH decreases by 1 unit, the hydrogen ion concentration **increases by a factor of 10 or 1 order of magnitude**. On the other hand, if the pH increases by 1 then the hydrogen ion concentration decreases by a factor of 10 or has decreased by 1 order of magnitude to one-tenth of what it was before.

What happens to the pH as the concentration of H^+ ions goes from 0.1 mol per dm^3 to 0.001 mol/dm^3?

In going from 0.1 mol/dm^3 to 0.001 mol/dm^3 the concentration decreases 100 times.

A decrease in the hydrogen ion concentration of 100 times is a decrease of 2 orders of magnitude and this means the pH value goes up by 2.

MATHS SKILLS

Understand what is meant by order of magnitude.

An increase of 1 order of magnitude means that the value has increased 10 times. A decrease by 1 order of magnitude means that the value has decreased 10 times to one-tenth of the original value.

An increase of 100 times is 2 orders of magnitude and so on.

CHECK IT! ✓

H 1 What is the difference between a strong acid and a weak acid?

H 2 What is the difference between a dilute and a concentrated solution?

H 3 If the pH of a solution goes down by 3 what has happened to the concentration of the H^+ ions in the solution?

The basics of electrolysis and the electrolysis of molten ionic compounds

Electrolysis is the splitting up of an ionic compound using electricity.

Electrolysis takes place when the ionic compounds are in the liquid state or in aqueous solution because the ions are free to move and carry the current.

The liquid that is decomposed by electrolysis is called the electrolyte.

When ions lose or gain electrons at the electrodes they are discharged.

During electrolysis, negative ions move towards the positive electrode (the anode (+)) and lose electrons to form non-metallic elements.

Positive ions move towards the negative electrode (cathode (−)) and gain electrons to form metallic elements (or hydrogen).

For example, molten potassium iodide forms iodine at the anode (+) and potassium at the cathode (−).

The reactions at electrodes can be represented using ionic half equations:

Generally at cathode (−) $M^{n+} + ne^- \rightarrow M$

At anode (+) $2X^{m-} \rightarrow X_2 + 2me^-$

These equations show that at the cathode the metal ions gain electrons and this is reduction. Also, at the anode the non-metal ions lose electrons and this is oxidation.

DOIT!

Write a short account of what happens to the ions and the electrons at each electrode during electrolysis. You could print out the diagram in the Snap it! box and complete it by putting in the ions and what happens at each electrode.

SNAPIT!

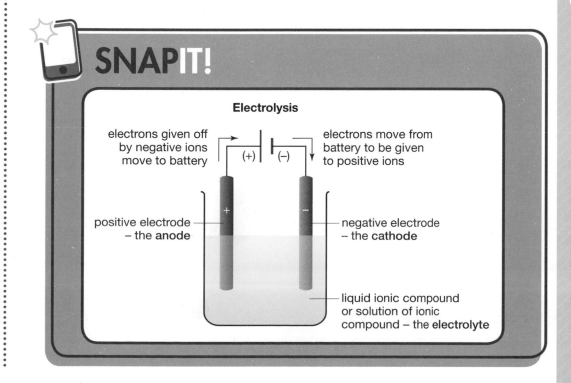

Electrolysis

electrons given off by negative ions move to battery

electrons move from battery to be given to positive ions

(+)   (−)

positive electrode – the **anode**

negative electrode – the **cathode**

liquid ionic compound or solution of ionic compound – the **electrolyte**

NAIL IT!

If you are doing the Higher Tier exam, then you should learn how to balance ionic half equations.

Look back to your notes on ionic compounds – remember that Group 1 elements form +1 ions, e.g. Na^+; Group 2 elements form +2 ions: Group 6 elements form 2– ions; and Group 7 elements form 1– ions.

MATHS SKILLS

You need to balance the charges on both sides of the equation.

WORK IT!

Foundation Tier What is formed at the cathode (–) and anode (+) when molten calcium chloride is electrolysed?

Calcium is formed at the cathode and chlorine at the anode.

Higher Tier   Give the equations for the reactions at the electrodes.

At the cathode (–) $Ca^{2+} + 2e^- \rightarrow Ca$

At the anode (+) $2Cl^- \rightarrow Cl_2 + 2e^-$

Flakes of calcium chloride

CHECK IT! ✓

1 What is the term used for the following:

   a The liquid that is electrolysed

   b The negative electrode

   c The positive electrode?

2 When molten magnesium chloride is electrolysed what substances are formed at

   a the anode

   b the cathode?

H 3 a Write ionic equations for the reactions taking place at both electrodes when molten magnesium chloride is electrolysed.

   b Explain why the reaction at the anode is oxidation.

The electrolysis of aqueous solutions

Water ionises to give hydrogen ions (H$^+$(aq)) and hydroxide ions (OH$^-$(aq)).

At the cathode ($-$), if the H$^+$(aq) ion is discharged then hydrogen (H$_2$) gas is produced.

At the anode ($+$), if the hydroxide ion is discharged then oxygen is produced.

The presence of these two ions from the water means that there is a choice from two products at each electrode.

At the cathode we have a choice between the metal ion and the hydrogen ion.

If the metal present is more reactive than hydrogen, then we get **hydrogen from the discharge of the hydrogen ions**.

At the anode ($+$) we have a choice between the discharge of the hydroxide ion and the other negative ion.

We get oxygen from the discharge of the hydroxide ion unless the other negative ion is a Group 7 halide ion (Cl$^-$, Br$^-$ or I$^-$). If so we get the halogen which is Cl$_2$, Br$_2$ or I$_2$.

DOIT!

Choose a solution. Draw a diagram showing both electrodes and the ions that would be attracted to those electrodes. Write notes on your diagram to explain what happens to the ions.

WORKIT!

Electrolysis of aqueous solutions

1   How can we predict which positive ion is discharged at the cathode ($-$) during the electrolysis of an aqueous solution?

The answer is hydrogen unless the metal is less reactive than hydrogen. This means that in your reactivity series list (PoSLiCaMZIC) only copper ions (C in the list) would be discharged.

Cu2+ ions accept electrons to give Cu.

2   How can we predict which non-metal ion is discharged at the anode during electrolysis of an aqueous solution?

If the ion other than hydroxide (OH$^-$) is a halide ion (Cl$^-$, Br$^-$ or I$^-$) then the halide ion is discharged to give the halogen Cl$_2$, Br$_2$ or I$_2$. If there is any other negative ion (such as SO$_4$$^{2-}$ and NO$_3$$^-$) then we get oxygen given at the anode from the discharge of the hydroxide (OH$^-$) ion.

NAILIT!

REMINDER – The product at the anode is always a diatomic molecule. The elements forming diatomic molecules are H_2, O_2, N_2, Cl_2, Br_2, I_2 and F_2.

WORKIT!

When aqueous potassium chloride solution is electrolysed, what is formed at each electrode and what solution remains after the electrolysis?

Potassium is more reactive than hydrogen and this means that at the cathode (−) we will get hydrogen (H_2) as the product.

Chloride (Cl^-) is a halide ion so we will get chlorine (Cl_2) formed at the anode (+).

This means that the ions left behind are potassium (K^+) and hydroxide (OH^-) ions and the solution that remains is potassium hydroxide (KOH) solution.

 **Practical Skills**

A required practical is to investigate what happens when aqueous solutions are electrolysed using inert electrodes.

STRETCHIT!

At the Higher Tier you are asked to write half-equations for the reactions at the electrodes. So when aqueous solutions are electrolysed you may be asked to write the half equations for the discharge of the hydrogen (H^+) and the hydroxide (OH^-) ions.

The half equations for the reactions at each electrode are shown below.

At the negative cathode (−) $2H^+(aq) + 2e^- \rightarrow H_2(g)$

And at the positive anode (+) $4OH^-(aq) \rightarrow 2H_2O(l) + O_2(g) + 4e^-$

This means that at the cathode, hydrogen ions gain electrons and are reduced. At the same time at the anode, hydroxide ions lose electrons and are oxidised.

NAILIT!   H

When asked what products you would get from the electrolysis of an aqueous solution you should also be aware of what remains. For example, if you have a solution of copper(II) sulfate you will get copper at the cathode because copper is less reactive than hydrogen. At the anode you get oxygen from the hydroxide ion. This means that hydrogen ions and sulfate ions are left behind, which together make sulfuric acid.

At the Higher Tier you may be asked to write the half-equations for the reactions at the electrodes.

CHECKIT! ✓

1 **a** Give the formulae of the four ions present in an aqueous solution of sodium chloride.

   b What are the products at each electrode when aqueous sodium chloride solution is electrolysed?

   c What solution remains after the electrolysis?

   H d Write the half equations for the reactions at each electrode and **explain** whether oxidation or reduction has taken place.

The extraction of metals using electrolysis

If a metal is more reactive than carbon then the metal oxide cannot be reduced to the metal by heating with carbon.

Electrolysis is used to extract metals **more reactive** than carbon. An example is aluminium.

To extract aluminium the electrolyte used is aluminium oxide. Aluminium oxide has a very high melting point so to save energy and lower the operating temperature the aluminium oxide is dissolved in a compound called cryolite.

Carbon is used for both electrodes. Aluminium ions are discharged at the cathode (–) to give molten aluminium metal which is run off.

At the anode (+) oxide ions are discharged to form oxygen gas. This oxygen gas then reacts with the carbon anode to form carbon dioxide. The carbon anode (+) burns away and loses mass. This means that the anode (+) has to be replaced at regular intervals.

DOIT!

Write a brief description of how aluminium is extracted from aluminium oxide. Your description should include materials used for electrodes, how energy is saved and what happens at each electrode.

SNAPIT!

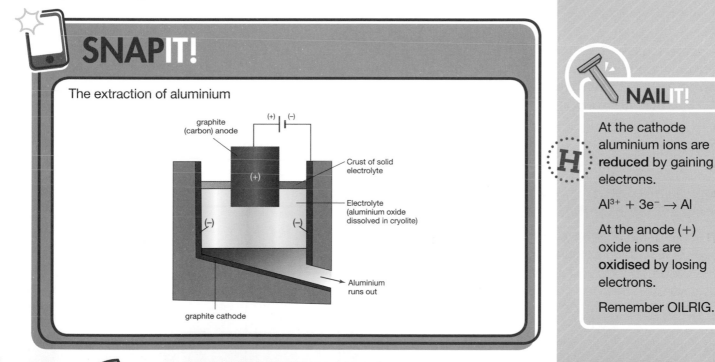

The extraction of aluminium

graphite (carbon) anode
(+) (–)
Crust of solid electrolyte
(+)
Electrolyte (aluminium oxide dissolved in cryolite)
(–)   (–)
Aluminium runs out
graphite cathode

NAILIT!

H At the cathode aluminium ions are **reduced** by gaining electrons.

$$Al^{3+} + 3e^- \rightarrow Al$$

At the anode (+) oxide ions are **oxidised** by losing electrons.

Remember OILRIG.

CHECKIT!

1 List the metals that are extracted using electrolysis.

2 Why is aluminium extracted using electrolysis?

3 Write a brief description of the electrolysis of aluminium oxide.

H 4 Sodium is extracted from molten sodium chloride. Write the half equations for the reactions at each electrode.

Practical: Investigation of the electrolysis of aqueous solutions

Part of the investigation is to predict the products formed by the electrolysis of various solutions. See page 75.

You can make predictions about the identity of the solution that remains because you know that it is formed from the ions that do not react.

Universal indicator can be used to identify the pH of the solution.

The apparatus you will use is similar to that shown in the Snap it! box. The gas is collected by downward displacement of the solution used.

The volume of gas produced depends on the identity of the gas. If hydrogen and oxygen are formed, then the volume of hydrogen is twice that of the volume of oxygen.

The volume of chlorine gas is less than predicted because chlorine dissolves in water.

The products at the electrodes are identified by chemical tests:

Hydrogen gas 'pops' when a lighted splint is placed in the gas.

Oxygen gas relights a glowing splint.

Chlorine bleaches blue litmus paper or UI in the solution.

Bromine turns the solution yellow/orange and iodine will turn it brown.

Practical Skills

This practical tests the skills of planning and predicting, carrying out, making observations and analysing the results.

DO IT!

Describe how the apparatus shown in the Snap it! box is used in the investigation.

SNAPIT!

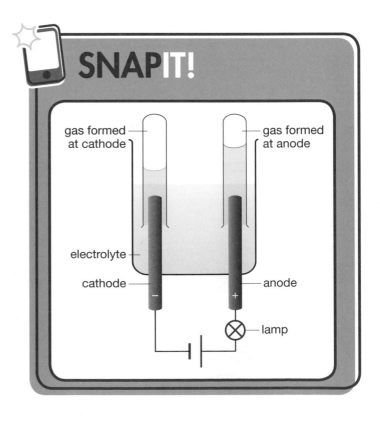

WORKIT!

If an aqueous solution of sodium chloride is electrolysed what would be the predictions and what would be seen?

Predictions			Observations		
At cathode	At anode	Solution	At cathode	At anode	Solution
Hydrogen formed because sodium is more reactive than hydrogen.	Chloride is a halide ion so chlorine would be formed from the discharge of the chloride ion.	The ions left behind are sodium and hydroxide ions.	The gas pops with a lighted splint which shows hydrogen.	The gas formed bleaches the UI in the solution. This shows chlorine is formed.	The remaining solution turns UI purple around the cathode. This is because the ions remaining are sodium and hydroxide ions which form the alkali sodium hydroxide.

In practicals make sure you record your observations!

NAILIT!

There are several areas of chemical knowledge that are used in this practical so make sure you are able to answer questions on them.

1 The electrolysis of aqueous solutions.

2 Testing for gases.

3 Formulae of ions.

4 The colour of UI in different types of solution.

CHECKIT!

1 In an investigation the gas at the cathode popped with a lighted splint and the gas at the anode relit a glowing splint. Name these two gases.

2 Explain why the solution around the anode turns yellowy orange when aqueous potassium bromide is electrolysed.

3 When sodium iodide solution is electrolysed the following reaction takes place:

$H_2O(l) + 2NaI(aq) \rightarrow H_2(g) + I_2(aq) + 2NaOH(aq)$

The iodine turns the solution brown. How would you identify the hydrogen and the sodium hydroxide solution?

Practical: Determining reacting volumes by titration

When acids are neutralised by alkalis there are no visible changes.

To see when the neutralisation is complete an indicator is added to the reaction mixture. The indicator **changes colour** when the correct volumes of acid and alkali have reacted.

Accurate volumes of acid and alkali are measured using a burette and a pipette.

Volumes are expressed to 2 decimal places.

SNAPIT!

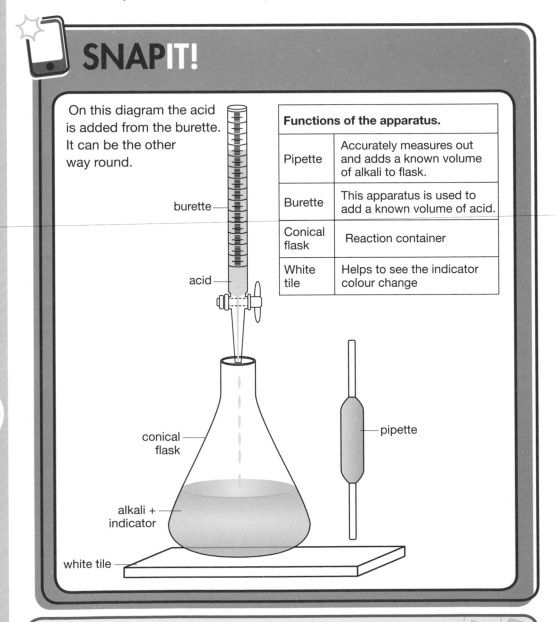

On this diagram the acid is added from the burette. It can be the other way round.

burette

acid

conical flask

alkali + indicator

white tile

pipette

Functions of the apparatus.	
Pipette	Accurately measures out and adds a known volume of alkali to flask.
Burette	This apparatus is used to add a known volume of acid.
Conical flask	Reaction container
White tile	Helps to see the indicator colour change

NAILIT!

The burette can be read to the nearest 0.05 cm³ so the results are expressed to 2 decimal places. Please note that zero is recorded as 0.00 not 0 because the burette reads to 0.05 cm³ so results are recorded to 2 decimal places.

MATHS SKILLS

You may have to read a table of data and use simple mathematical operations such as subtraction and finding the mean of a set of results. Results are expressed to 2 decimal places.

WORKIT!

Given below is a table of results from a titration experiment where an acid is added from the burette to 25.00 cm³ of an alkali.

There are two results that are within 0.10 cm³ of each other and the average of these is used as the reading.

What is the average titration result for this experiment?

Reading	Rough titration	Accurate titration 1	Accurate titration 2	Accurate titration 3
Final reading/cm³	19.00	37.20	18.50	37.10
Initial reading/cm³	0.00	19.00	0.00	18.50
Titre/cm³	19.00	18.20	18.50	18.60

The average accurate titre = (18.50 + 18.60)/2 = 18.55 cm³

Practical Skills

Titration procedure

The point at which the indicator changes colour is the end-point. The volume of acid needed to reach the end-point is called the titre.

For each titration a known volume of alkali is added to the conical flask using the pipette, followed by a couple of drops of indicator.

The first titration is a rough titration – the acid is added from the burette 1 cm³ at a time until the end-point. This gives an estimate of the volume of acid required in the reaction.

Using the rough titration reading, acid is run into alkali until 1.00 cm³ before the end-point for the rough titration and then add the solution drop-by-drop to give an accurate titre.

This is repeated until you get two results which are within 0.10 cm³ of each other.

After the rough titration it is good practice to start the next titration where you stopped last time. This means that in the example above we start at 19.00 cm³ and not at 0.00. This saves time and materials.

Note that the recording of results is a very important part of titrations. You need a column for your rough titration and columns for each accurate titration.

✓ CHECKIT!

1 Name the main apparatus used for a titration.

2 What is a rough titration?

3 How can you tell when enough acid has been added to the alkali?

4 In a titration experiment four accurate results were obtained: 24.20 cm³, 24.50 cm³, 24.60 cm³ and 25.80 cm³.

   a Which results are rejected?

   b What is the average titre for the experiment?

For additional questions, vi
www.scholastic.co.uk/g

1 The five metals aluminium, copper, iron, magnesium and zinc are in the reactivity series.

   a Put them in order of reactivity from the **least reactive** to **the most reactive**.

   b Complete the following word equations for each reaction :

     i   zinc(s) + copper(II) sulfate(aq) →

     ii aluminium(s) + magnesium oxide(s) →

     iii aluminium(s) + iron(III) oxide(s) →

   c When burning magnesium is lowered into a test tube containing carbon dioxide, the magnesium continues burning.

     At the end of the reaction there are traces of a black solid and a white solid on the sides of the test-tube.

     i   Write the word equation for the reaction including state symbols.

     ii Write the balanced symbol equation for the reaction including state symbols.

     iii Explain why this is a redox reaction.

     iv I Identify the white solid formed.

         II Identify the black solid formed.

2 When hydrochloric acid solution is added to zinc metal, the zinc disappears and the mixture effervesces but when the same acid is added to copper metal there is no reaction.

   a Explain the effervescence and describe how you can test for the product that causes the effervescence.

   b Explain why there is no reaction between the acid and copper.

   H c i Write the ionic equation for the reaction between zinc metal and the acid. Note that the formula for the zinc ion is Zn^{2+}.

     iii Explain why this is a redox reaction.

3 In the electrolysis of aqueous potassium bromide solution there is a gas produced at the cathode and at the anode the solution turns yellow-orange in colour.

   a Name the products produced at both electrodes.

   b What solution remains after the electrolysis?

4 a A solution of ammonia turns universal indicator blue. What does this tell you about the ammonia solution?

   b Universal indicator turns yellow in phenol solution. What does this tell you about phenol?

5 A student was given a solution of hydrochloric acid and a solution of sodium hydroxide. He was asked to carry out a titration to find the volume of hydrochloric acid that would exactly react with $25.00\,cm^3$ of the sodium hydroxide solution.

   a The equation for the reaction is:

     sodium hydroxide(aq) + hydrochloric acid(aq) → sodium chloride(aq) + water(l)

     i   What is the meaning of (aq)?

     ii Write the balanced symbol equation for the reaction.

     iii Write the ionic equation for the reaction.

   b In the titration what equipment does he use for the following:

     i   Measuring exactly $25.00\,cm^3$ of the sodium hydroxide solution

     ii Measuring and adding the acid to the sodium hydroxide solution

     iii The reaction vessel?

   c How can he tell when the reaction has finished?

Energy changes

Exothermic and endothermic reactions

An exothermic reaction is one where the reaction gives out heat energy to its surroundings. This results in an increase in the temperature of the surroundings.

Examples of exothermic reactions are the burning of fuels in combustion reactions and most neutralisation and oxidation reactions.

Everyday uses of exothermic reactions are hand-warmers and self-heating cans of food.

An endothermic reaction is one where the reaction takes in heat energy from its surroundings. This results in a decrease in the temperature of the surroundings.

Examples of endothermic reactions are thermal decomposition (breaking up a compound using heat) and the reaction of citric acid with sodium hydrogen carbonate.

Sports injury packs use endothermic reactions.

SNAPIT!

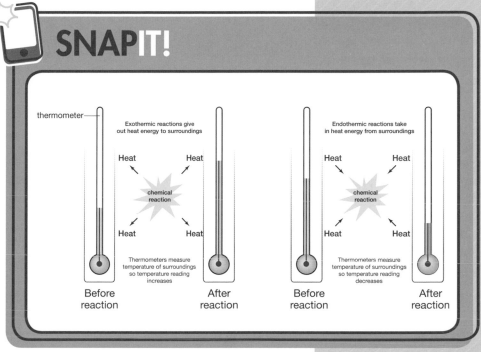

thermometer

Exothermic reactions give out heat energy to surroundings

Heat   Heat

chemical reaction

Heat   Heat

Thermometers measure temperature of surroundings so temperature reading increases

Before reaction   After reaction

Endothermic reactions take in heat energy from surroundings

Heat   Heat

chemical reaction

Heat   Heat

Thermometers measure temperature of surroundings so temperature reading decreases

Before reaction   After reaction

DOIT!

Write a brief summary of the diagram in the Snap it! box.

What do the changes in the thermometers mean?

CHECKIT!

1 a If the temperature goes up in a chemical reaction, what type of reaction is taking place?

   b The table below shows the results from two experiments:

Experiment	Initial temperature/°C	Final temperature/°C	Temperature change
I	20		+25
II	20	15	

Complete the table and then classify each reaction as either exothermic or endothermic.

2 List some types of reaction that are endothermic reactions.

3 List some uses of exothermic reactions.

Practical: Investigation into the variables that affect temperature changes in chemical reactions

This is one of the required practicals and you could be questioned on it in the exam. This particular practical places emphasis on you putting forward a hypothesis and making predictions based on it. It also requires the accurate use of appropriate measuring apparatus and the need to make and record a range of measurements. When investigating temperature changes you should know what factors need to be controlled to make a fair test.

Exothermic reactions are accompanied by a temperature rise and **endothermic** reactions give a temperature decrease.

Reactions you could investigate include combustion, neutralisation reactions between acids and alkalis, and acids reacting with either metals or carbonates.

Factors affecting the temperature changes are amount of reactant, surface area (lumps or powder) of solids, concentrations of solutions and the reactivity of different metals with acids.

Relevant pieces of apparatus are thermometers or temperature data loggers; measuring cylinders; spirit burners for combustion experiments – see diagram below; top-pan balance and reaction containers such as beakers and test tubes. The container that is used to measure the heat change for a reaction is called a calorimeter. A polystyrene cup can be used or a lagged container like a beaker. The lagging reduces heat loss through the sides of the container.

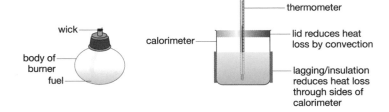

DO IT!

When you are carrying out an investigation it is often useful to draw a flow chart of what you are going to do. Discuss it with a practical partner or review it yourself to identify where you might be doing things in the wrong order or points where a piece of apparatus is needed. Practical work is very much improved by good organisation.

Practical Skills

One required practical is to investigate the variables that affect temperature changes in reacting solutions.

For this practical you should:

* Put forward a hypothesis about what you think would happen based on your knowledge of the chemistry involved.
* Identify variables that could affect the temperature change and explain how you would make it a fair test.
* Identify safety factors involved in the experiment.
* Know what measurements you would make.
* Record your results in a suitable form.

STRETCH IT!

If you are taking the Higher Tier exam then you should be aware that if you are investigating different substances then equal masses does not mean equal amounts, and masses should be converted to moles using n = m/M_r.

WORKIT!

A student was given some magnesium ribbon, magnesium powder and hydrochloric acid. She was asked to investigate the effect of the surface area of the magnesium on the temperature change caused by the reaction. She carried out a fair test and measured the starting temperature for both as 20°C. In the reaction with the magnesium ribbon the final temperature was 28°C and with the powder it was 42°C.

a List the apparatus she would need and give the use of each piece.

She would need the following apparatus:

- A small beaker for the reaction container which is very well insulated and a thermometer (or temperature data logger) to measure the temperature changes.
- A measuring cylinder to measure the volume of acid.
- A spatula for adding the powder and a top-pan balance for weighing out the magnesium ribbon and powder.

b Describe how she would make it a fair test.

The volume and concentration of the acid should be the same for both experiments and so should the mass of the ribbon and powder. The containers should be identical or use the same one for both experiments.

c Show how she might record her results.

Experiment	Starting temperature/°C	Final temperature/°C	Temperature change/°C
Magnesium powder	20	42	22
Magnesium ribbon	20	28	8

d Make any conclusions possible from her results.

The reaction with the powder gave a greater temperature rise. This is because it had the greater surface area and reacted more quickly.

CHECKIT!

NAILIT!

A common error when drawing up a table of results is to forget to put in the units.

1 List some factors that could affect the temperature changes in a chemical reaction.

2 A student added measured amounts of magnesium powder to separate calorimeters containing $100\,cm^3$ of $2\,mol/dm^3$ hydrochloric acid and then measured the temperature changes. The results are shown below:

Experiment	1	2	3	4	5	6
Mass of magnesium/g	1.00	2.00	3.00	4.00	5.00	6.00
Temperature change/°C	5.00	10.0	15.0	20.0	24.0	24.0

a Explain the results obtained for experiments 1 to 4.

b Suggest an explanation for the results from experiments 5 and 6.

3 List some reactions that could be investigated in terms of temperature changes.

4 Using their reactions with hydrochloric acid describe how you could place the metals copper, iron, magnesium and zinc in order of reactivity.

Reaction profiles

DO IT!

Draw reaction profiles for:

- A combustion reaction.
- A thermal decomposition.

NAIL IT!

The arrow for the activation energy must **start at the energy of the reactants and end at the peak of the reaction profile**. The line showing the change in energy of the reaction starts at the energy for the reactants and ends at the energy of the products. It **points downwards for an exothermic reaction** and **upwards for an endothermic reaction.**

A reaction profile shows how the energy changes from reactants to products.

In a reaction profile for an exothermic reaction the products are lower in energy than the reactants because energy is released to the surroundings during the reaction.

In a reaction profile for an endothermic reaction the products are higher in energy than the reactants because energy is taken in from the surroundings during the reaction.

Chemical reactions occur when reacting particles collide with enough energy to react. This energy is called the activation energy (E_a).

The activation energy is the minimum energy required for a reaction to occur. The activation energy is a barrier to reaction.

SNAP IT!

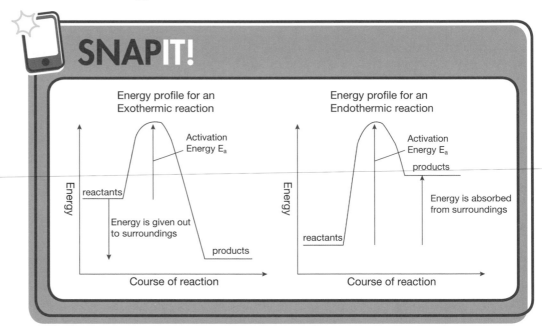

CHECK IT!

1 Draw and label the reaction profile for the combustion of carbon to give carbon dioxide.

2 a i Refer to page 99 and define what is meant by a catalyst.

   ii Explain how a catalyst speeds up a reaction.

 b i Draw a reaction profile for an endothermic reaction.

   ii On the same graph, draw the profile you would get for a catalyst for the same endothermic reaction.

The energy changes of reactions

Energy is absorbed when chemical bonds are broken in a chemical reaction.

Energy is given out when chemical bonds are formed in a chemical reaction.

Bond energies are used to calculate the energy needed to break bonds and the energy given out by making bonds.

The energy given out or taken in during a chemical reaction is measured in kJ/mol.

The energy difference between the energy needed to break the bonds and the energy given out by their formation is the energy change in the reaction.

If the energy needed to break the bonds in the reactants is less than the energy given out by the formation of new bonds in the products, then the reaction is exothermic.

If the energy needed to break the bonds in the reactants is greater than the energy given out by the formation of new bonds in the products, then the reaction is endothermic.

DOIT!

Methane and oxygen react as follows:

$$CH_4(g) + 2O_2(g) \rightarrow CO_2(g) + 2H_2O(l)$$

NAILIT!

It is a common error to miscalculate the number of bonds in a molecule. For example in the Snap it! box there are three carbons in the propane (C_3H_8) molecule and it is tempting to put this as 3 C-C bonds when in fact there are only 2 C-C bonds (C-C-C). Also remember that in a water molecule there are 2 O-H bonds and in carbon dioxide there are 2 C=O bonds.

DOIT!

For the reaction between methane and oxygen draw a reaction profile and on it show the bonds broken in the reactants and the bonds formed in the products.

The burning of propane in oxygen to give water and carbon dioxide:

$$C_3H_8 \quad + \quad 5O_2 \quad \longrightarrow \quad 3CO_2 \quad + \quad 4H_2O$$

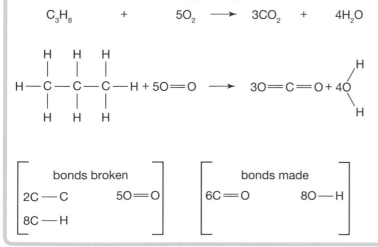

bonds broken		bonds made	
2C — C	5O=O	6C=O	8O — H
8C — H			

MATHS SKILLS

You will need to read data from a table and carry out simple multiplications, additions and subtractions.

WORKIT!

Ⓗ

Calculate the energy change for the combustion of propane shown in the equation below. The required bond-energies are shown in the table.

You will be supplied with this information in the exam.

$$C_3H_8 \quad + \quad 5O_2 \quad \longrightarrow \quad 3CO_2 \quad + \quad 4H_2O$$

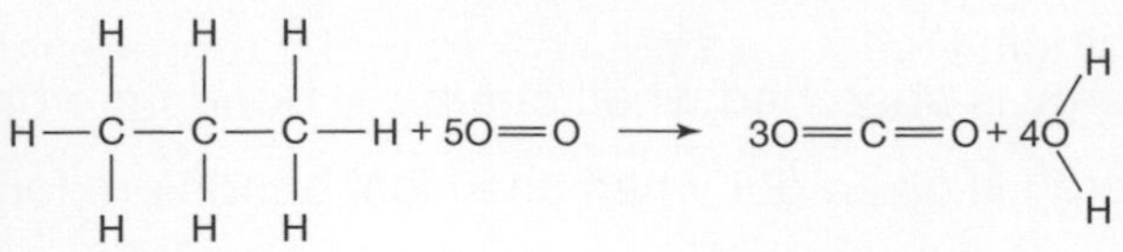

Bond	Bond energy (kJ per mol)
C–C	350
C–H	415
O=O	500
C=O	800
O–H	465

H—C—C—C—H + 5O=O ⟶ 3O=C=O + 4O⟨H H

bonds broken
2C—C   5O=O
8C—H

bonds made
6C=O   8O—H

Bonds broken: heat energy taken in	Bonds made: heat energy given out
2 C–C bonds need 2 × 350 kJ = 700 kJ of energy to break them	6 C=O bonds give out 6 × 800 kJ of energy when they form = 4800 kJ
8 C–H bonds need 8 × 415 = 3320 kJ of energy to break them	8 O–H bonds give out 8 × 465 kJ of energy when they are formed = 3720 kJ
5O = O bonds need 5 × 500 = 2500 kJ of energy to break them	
Total energy taken in to break bonds = 700 + 3320 + 2500 kJ = 6520 kJ	Total energy given out when bonds are made = 4800 + 3720 kJ = 8520 kJ

There is more energy given out than taken in so the reaction is exothermic.
The energy of reaction = 8520 − 6520 = 2000 kJ/mol

CHECKIT! ✓

H 1 When methane reacts with oxygen the following reaction takes place:

$$CH_4(g) + 2O_2(g) \rightarrow 2H_2O(l) + CO_2(g)$$

a Draw the bonds present in the molecules.

b Use the values in the table above to calculate:

   i the energy taken in to break bonds

   ii the energy given out when bonds are formed

   iii the energy change for the reaction.

c Is the reaction exothermic or endothermic? Explain your answer.

Chemical cells and fuel cells

If you place two different metals in an electrolyte (a liquid that conducts electricity) electrons will flow from the more reactive metal to the less reactive metal.

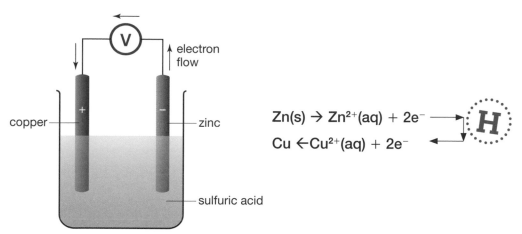

$$Zn(s) \rightarrow Zn^{2+}(aq) + 2e^-$$
$$Cu \leftarrow Cu^{2+}(aq) + 2e^-$$

DO IT!

You are given a copper plate, a chromium plate, a magnesium plate, zinc plate, some sulfuric acid, beakers and a voltmeter. Devise an experiment that helps you put the chromium, magnesium and zinc in order of reactivity.

For example, if a zinc plate and a copper plate are placed in sulfuric acid (the electrolyte), the more reactive zinc loses electrons more easily than the copper. If the two plates are connected then these electrons can then flow from the zinc to the copper.

This electron flow can be used to power an electric circuit.

The greater the difference in reactivity of the two metals the greater the voltage produced by the cell.

Cells contain chemicals that undergo redox reactions to produce electricity.

When two or more cells are connected in series they form a battery and the voltage produced goes up as the number of cells in the battery increases.

There are two types of batteries – rechargeable and non-rechargeable.

In non-rechargeable batteries such as alkaline batteries, the chemical reactions that produce the electricity stop when one of the reactants is used up.

In hydrogen fuel cells the reaction between hydrogen and oxygen produces electrical energy rather than heat energy. The only product of the fuel cell reaction is water.

The cell is powered by a constant flow of hydrogen and oxygen and as long as these gases flow the fuel cell will never run out, unlike non-rechargeable cells.

H
At the Higher Tier the two half equations for the electrode reactions are required (see Stretch It! box). **H**

89

STRETCH IT!

At the negative electrode of the **fuel cell** hydrogen reacts with hydroxide ions to produce water and electrons.

$2H_2(g) + 4OH^-(aq) \rightarrow 4H_2O(l) + 4e^-$

This is oxidation because the electrons are lost.

At the positive electrode oxygen gains electrons and reacts with water to produce hydroxide ions.

$O_2(g) + 2H_2O(l) + 4e^- \rightarrow 4OH^-(aq)$

This is reduction because electrons are gained.

When the left-hand side and the right-hand side of these 2 reactions are added together and we cancel out we get:

$2H_2(g) + 4OH^-(aq) + O_2(g) + 2H_2O(l) + 4e^-$
$\rightarrow 4OH^-(aq) + 4H_2O(l) + 4e^-$

Which becomes $2H_2(g) + O_2(g) \rightarrow 2H_2O(l)$

SNAP IT!

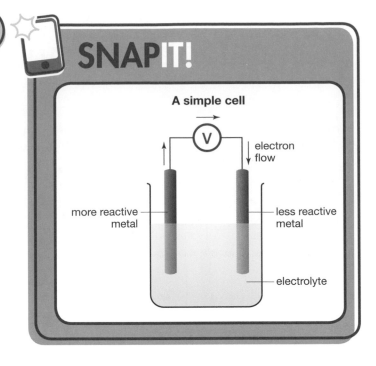

A simple cell

electron flow

more reactive metal — less reactive metal

electrolyte

NAIL IT!

You could be asked about the relative advantages and disadvantages of chemical cells and fuel cells.

Chemical cells	Fuel cells
Can be used anywhere.	Hampered by the need for hydrogen containers.
When non-rechargeable batteries run out, they have to be thrown away and sent to a recycling centre. Rechargeable batteries can be charged again and again.	These will continue to work as long as the hydrogen and oxygen flows. The product of the reaction is water.
Some of the metals used are toxic.	The hydrogen is flammable.

CHECK IT!

1 Describe the basic requirements for making a chemical cell.

2 What is the main product of a fuel cell?

3 Explain which of the following two combinations would give the greater voltage:

A cell consisting of magnesium and copper electrodes OR one consisting of zinc and copper electrodes.

H 4 State the two half equations for the electrode equations of a fuel cell.

H 5 a Write the half equations for the reactions taking place in the cell set up by placing magnesium and zinc plates in an electrolyte of sulfuric acid.

b Draw a diagram of this cell and show the direction of the electron flow.

1 What happens to the temperature in a reaction when the reaction is endothermic?

2 When limestone is heated it undergoes thermal decomposition to give calcium oxide (known as quicklime) and carbon dioxide. The reaction between calcium oxide and water can be used to heat up meals. The relevant equations are shown below:

A $CaCO_3(s) \rightarrow CaO(s) + CO_2(g)$

B $CaO(s) + H_2O(l) \rightarrow Ca(OH)_2(s)$

Classify reactions A and B as either exothermic or endothermic. In each case explain your answer.

3 a What are the main differences between a chemical non-rechargeable cell and a hydrogen fuel cell?

b A cell was made from a copper rod and a zinc rod which were placed in an electrolyte. Explain which rod the electrons would flow from if the cell was placed in an electrical circuit.

4 The **incomplete** table below shows the results from an experiment on the reactivity of three metals, X, Y and Z. The same amounts of all three metals were reacted with hydrochloric acid and the temperature change in each reaction was measured.

Metal	Starting temperature	Final temperature	Temperature change
X	22		1
Y	22		19
Z	22	29	

a i Complete the table.

ii Apart from the missing temperature figures what else is missing from the table?

b How could you make sure that the comparison is a fair one?

c What apparatus would you use for this experiment?

d Give the order of reactivity of the three metals and explain your answer.

5 The diagram opposite shows the reaction profile for a chemical reaction.

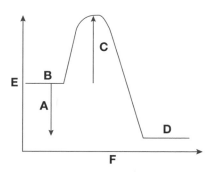

a Write down the correct labels for A to F.

b What type of reaction is represented by this profile?

c Explain what is meant by the term 'activation energy'.

H 6 The equation below shows the structures of the reactants and products of the chemical reaction between ethene and chlorine.

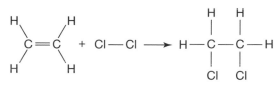

ethene   chlorine   dichloroethane

a Show the bonds that are broken and calculate the energy required to break all these bonds.

b Show the bonds that are made and calculate the energy given out by their formation.

c i Calculate the energy change for the reaction and the units.

ii Explain whether it is exothermic or endothermic.

Table of bond energies

All values are in kJ per mol.

Bond	Bond energy
C=C	610
C–H	415
Cl–Cl	245
C–C	350
C–Cl	345

Rates of reaction and equilibrium

Ways to follow a chemical reaction

The main ways of following a chemical reaction to measure its rate are:

- measuring the volume of gas produced over a period of time
- measuring the loss in mass of the reactants over a period of time
- measuring how long it takes for a cross to be obscured when a solid is formed in a reaction between two solutions.

When you investigate the factors that affect the rate of a chemical reaction there are several variables that can be changed, measured or controlled. For example, if you were investigating the effect of temperature on the rate of reaction between marble chips and acid, you could measure the volume of gas produced over time. To make it a fair test you would keep the concentration of acid and the surface area of the marble chips constant for all the experiments.

The table below summarises facts about variables using this example.

Type of variable	Description	In the example it is....	Where it is plotted on a graph
Independent	The variable whose effect you have chosen to measure	Temperature	On the horizontal or x-axis
Dependent	The variable that you measure to find the effect of changing the independent variable	The volume of gas	Up the vertical or y-axis
Control variables	The variables you keep constant to make it a fair test	Concentration of acid and surface area of the marble chips	These are not plotted but could be noted in the title of the graph

SNAPIT!

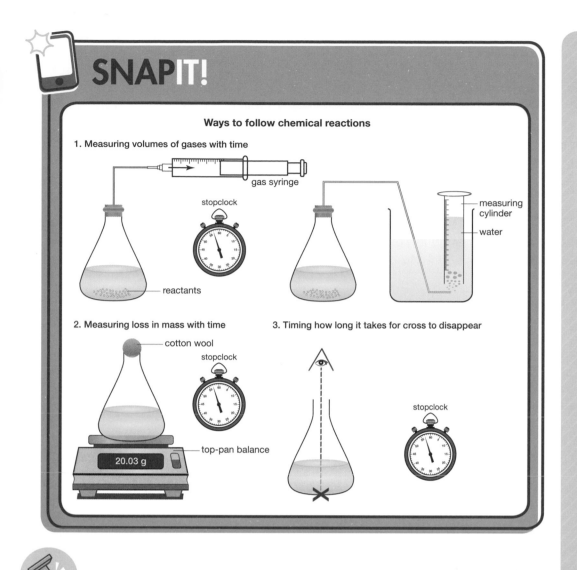

Ways to follow chemical reactions

1. Measuring volumes of gases with time

gas syringe

stopclock

reactants

measuring cylinder

water

2. Measuring loss in mass with time

cotton wool

stopclock

top-pan balance

20.03 g

3. Timing how long it takes for cross to disappear

stopclock

DOIT!

Using the diagrams in the Snap it! box describe briefly how you would make the measurements in each method and create a suitable results table for each experiment.

NAILIT!

If you are asked to pick an appropriate method for following a chemical reaction to see how quickly it is going, then the equation will help you.

You can use the state symbols in the equation to help you decide on the method to use. For example, if you see the (g) state symbol in the products, then you know a gas is produced.

If a gas is given off then measurement of gas volumes using the gas syringe or displacement of water are obvious alternatives. If the gas given off is carbon dioxide then measuring the loss in mass is also a possibility.

When two solutions react to form a solid then the mixture goes cloudy and you can time how long it takes to obscure a cross.

If a solid disappears during a reaction then you can time how long it takes for this to happen.

STRETCHIT!

If you use the disappearing cross technique then you measure how long it takes to obscure the cross so that you cannot see it.

The longer the time taken for this to happen, the slower the reaction is. If you want to have an idea of how quick the reaction is, you measure the time but use 1/time when you plot your results because 1/time tells you how quick the reaction is.

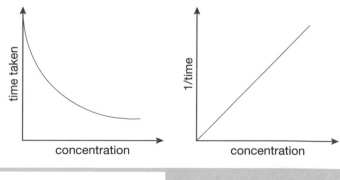

CHECKIT!

1 Which of the three methods shown on page 93 could be used to measure the rate of the following reactions? In some cases more than one method can be used.

   a $2HCl(aq) + CaCO_3(s) \rightarrow H_2O(l) + CaCl_2(aq) + CO_2(g)$

   b $2HCl(aq) + Mg(s) \rightarrow H_2(g) + MgCl_2(aq)$

2 A student was investigating the effect of concentration on the rate of the reaction between marble chips and hydrochloric acid. She decided to measure the volume of gas over a period of time.

   In this investigation what was:

   a the independent variable

   b the dependent variable?

3 The equation below shows the reaction between hydrochloric acid and sodium thiosulfate solution.

   $2HCl(aq) + Na_2S_2O_3(aq) \rightarrow SO_2(g) + 2NaCl(aq) + H_2O(l) + S(s)$

   The rate of this reaction is usually followed by timing how long it takes to obscure a cross.

   a Give some problems that might make this method inaccurate.

   b i You are given a small electric bulb, a power pack, some black card and a light data logger. Using a simple diagram explain how you could measure the rate of the same reaction using this apparatus.

      ii Explain how this method would be better than 'obscuring the cross' for measuring the rate.

You can measure the rate of reaction by measuring loss in mass with time

Calculating the rate of reaction

The rate of a chemical reaction can be looked at in two different ways. Either using the reactant being used up, in which case the

$$Mean\ rate\ of\ reaction = \frac{Amount\ of\ reactant\ used\ up}{Time\ taken}$$

Or in terms of the product formed in which case the

$$Mean\ rate\ of\ reaction = \frac{Amount\ of\ product\ formed}{Time\ taken}$$

When a graph is plotted to show the change in a product or reactant you will probably get a curve. The rate at any time can be found by drawing a tangent to the curve and measuring its slope.

The steeper the gradient of the tangent the faster the rate of reaction.

When the gradient=0 there is no reaction taking place and the reaction is complete.

DO IT!

Sketch a graph showing the course of a reaction and on it draw a few tangents. Write a brief description about why the tangents show that as time goes on the reaction gets slower.

WORKIT!

In this experiment the reaction is finished after 63 s and the total volume of gas collected is 50 cm³.

This means that the average rate of reaction = 50/63 = 0.794 cm³/s.

At X the tangent's gradient = 20/10 = 2 cm³/s.

At Y the tangent's gradient = 15/20 = 0.750 cm³/s.

At Z the gradient = 0 because the reaction has finished.

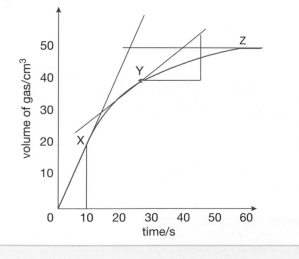

MATHS SKILLS

You will need to:

- Plot a graph or graphs and compare the results.

- Determine the slope of a linear graph.

- Understand that a tangent is a line that touches a curve at a point but does not cross it anywhere else.

- Draw tangents to a graph, calculate their gradients and use them as a measure of the rate of a reaction.

- Express results to three significant figures.

- Express very small or very large numbers in base form.

- Know that a straight line between concentration and rate shows that the rate is proportional to the concentration.

STRETCH IT!

You may be asked to express the rate in mol per s. Remember for a gas at RTP 1 mol of any gas occupies $24\,000\,cm^3$.

At **X** the rate $= 2\,cm^3/\,s$. $2\,cm^3 = 2/24\,000\,mol/s$

$\qquad\qquad = 8.33 \times 10^{-5}\,mol/s$

At **Y** the rate $= 0.750\,cm^3/s = 0.750/24\,000\,mol/s = 3.13 \times 10^{-5}\,mol/s$

NAIL IT!

For graph I the reaction giving graph A is faster than the reaction responsible for graph B, this is shown by the fact the line obtained is steeper at the beginning than the one representing reaction B.

Similarly for Graph II reaction C is faster than reaction D and this is why its line is steeper at the beginning than the line for reaction D. The line levels off as the reaction slows down and it is horizontal because the reaction has finished.

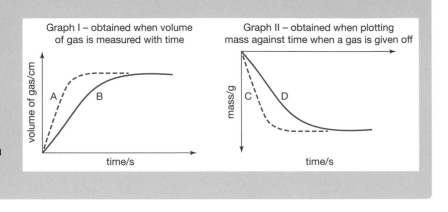

Graph I – obtained when volume of gas is measured with time

Graph II – obtained when plotting mass against time when a gas is given off

CHECK IT!

1 Describe how you can find the mean rate of a reaction using the amount of product formed in a reaction.

2 Two students carried out an investigation into the reaction between excess calcium carbonate and $100\,cm^3$ of hydrochloric acid.

$$CaCO_3(s) + 2HCl(aq) \rightarrow CaCl_2(aq) + H_2O(l) + CO_2(g)$$

They measured the volume of carbon dioxide produced at various time intervals. The results are shown in the table below.

Time/s	0	5	10	15	20	25	30	40	50	60	70	80
Volume of CO_2/cm^3	0	27	41	50	57	62	66	72	77	79	80	80

a Plot these results on a suitable graph.

b   i Draw tangents to the graph at 0 s, 10 s and 30 s.

   ii Find the rate of reaction in cm^3/s at each of these times.

   iii Explain why the rate decreases with time.

c Explain why the gradient of the tangent drawn at 80 s is zero.

H d   i How many moles of carbon dioxide were produced in the reaction?

   ii Calculate the number of moles of hydrochloric acid present.

   iii Calculate the concentration of the hydrochloric acid.

The effect of concentration on reaction rate and the effect of pressure on the rate of gaseous reactions

Particles have to collide in order to react.

When you increase the concentration of a solution in a chemical reaction the rate of the reaction also increases.

This is because when the concentration increases the reacting particles get more crowded as there are more of them in a given volume, so they collide more frequently and there are more successful collisions.

During a chemical reaction the concentration of the reactant particles in the solution decreases as they react. This means that they become less crowded and collide less **frequently** and the reaction slows down.

When a graph is plotted to show how a reaction is progressing, the slope of the tangent to the graph at any time shows us how quickly the reaction is at that time. The quicker the reaction the steeper the slope of the tangent.

Because the concentration of the reactant decreases during a reaction, the reaction gets slower and this means that the slope of the graph gets less steep and the graph becomes a curve.

In reactions that involve gases increasing the pressure also makes the reactant particles more crowded; they collide more frequently and react more quickly so the rate increases.

NAILIT!

When you talk about collisions saying just 'more collisions' will not get you the marks. There could be more collisions but over a much longer time. Using the word **frequently** is very important.

SNAPIT!

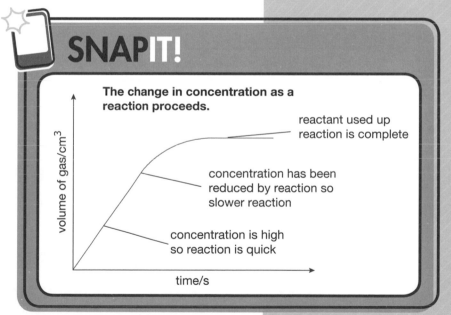

The change in concentration as a reaction proceeds.

volume of gas/cm³

reactant used up
reaction is complete

concentration has been reduced by reaction so slower reaction

concentration is high so reaction is quick

time/s

CHECKIT!

1 What happens to the rate if the concentration of a reacting solution decreases?

2 Using collision theory, explain why increasing the concentration has an effect on the rate.

3 When calcium carbonate reacts with excess hydrochloric acid, carbon dioxide gas is produced.

  a Sketch a graph of volume of carbon dioxide gas produced over time.

  b On the same scales sketch a graph showing the reaction using the same concentration of acid but only **half the amount** of calcium carbonate.

Rates of reaction – the effect of surface area

DOIT!

Imagine a cube of sides 1 cm. It has 6 faces. What is its total surface area? If the cube was broken up into smaller cubes with sides 0.1 cm what is the new surface area? Note that the number of smaller cubes is 1000.

If a solid is broken up into smaller pieces its surface area increases.

Smaller pieces have a larger surface area to volume ratio than larger pieces. This means that a powdered solid has a greater surface area than a lump.

If the experiment involves a comparison between the reaction rate given by a powdered reactant and a lump then a line graph cannot be drawn and the results are expressed using bar-charts.

A larger surface area means that more solid particles are exposed to collisions with other reactant particles because there are more points of contact.

This means that there are more frequent collisions and the rate of reaction increases.

So increasing the surface area of a solid reactant increases the rate of reaction.

SNAPIT!

Graphs to show effects of surface area

volume of gas/cm³ — lumps / powder

the gas is given off more quickly when the surface area is larger

time/s

time/s

loss in mass/g

The loss in mass is quicker when the surface area increases

lumps

powder

The steepness of the slopes of the graphs show how quickly the reaction is proceeding at any time. For both methods the initial (starting) slope of the graph is steeper for the powder showing that it is producing a faster reaction. Therefore the greater the surface area the faster the rate.

CHECKIT!

1 State which has the larger surface area - lumps or powder.

2 Explain in terms of collision theory what effect increasing the surface area has on the rate of reaction.

3 A student wanted to find the effect of surface area on the rate of the reaction between calcium carbonate (used as marble chips) and hydrochloric acid.

$CaCO_3(s) + 2HCl(aq) \rightarrow CaCl_2(aq) + H_2O(l) + CO_2(g)$

   a Explain how he could vary the surface area.

   b Give two ways he could follow the reaction rate.

   c How should he display his results?

The effects of changing the temperature and adding a catalyst

The activation energy is the minimum energy needed for a reaction to take place.

If colliding particles do not have an energy greater than the activation energy then they will not collide with enough energy to break bonds and react.

When the temperature is raised particles move around more quickly and have more kinetic energy.

This means that there are more frequent collisions because they collide more frequently (more collisions per second).

There are more frequent effective collisions because the particles have a greater chance of colliding with an energy greater than the activation energy.

Adding a catalyst to a reaction speeds up the reaction but it is unchanged chemically at the end of the reaction.

A catalyst lowers the activation energy for a reaction by providing an alternative pathway for the reaction which has a lower activation energy.

This means that there will be even more frequent effective collisions and therefore a faster rate of reaction.

Enzymes are biological catalysts and carry out reactions in living organisms.

The formulae for catalysts are not included in chemical equations because they are unchanged chemically by the reaction. They can be written on the arrow between the reactants and products, e.g.

hydrogen peroxide(aq) $\xrightarrow{\text{manganese dioxide catalyst}}$ water(l) + oxygen (g)

$2H_2O_2(aq) \xrightarrow{\text{manganese dioxide catalyst}} 2H_2O(l) + O_2(g)$

NAILIT!

For the variables concentration and surface area, doubling either of them will double the rate. This does not work with temperature. An approximate effect is that the rate for some reactions doubles if the temperature goes up by 10 °C.

DOIT!

Manganese dioxide is a catalyst for the decomposition of hydrogen peroxide to give water and oxygen gas.

You are given some hydrogen peroxide, manganese dioxide, wooden splints, test tubes and filtration apparatus. Explain how you could show that the manganese dioxide can be re-used as a catalyst.

SNAPIT!

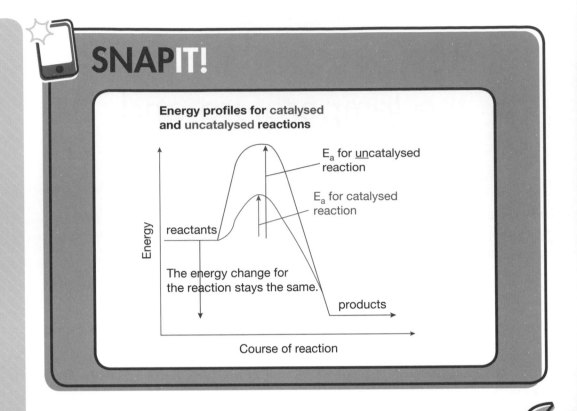

Energy profiles for catalysed and uncatalysed reactions

E_a for <u>un</u>catalysed reaction

E_a for catalysed reaction

reactants

The energy change for the reaction stays the same.

products

Course of reaction

CHECKIT!

1 What is the effect of increasing the temperature on the rate of reaction?

2 Using collision theory, explain the effect of raising the temperature on reaction rate.

3 a State what is meant by a catalyst.

   b How does a catalyst speed up a reaction?

4 The diagram below shows the energy levels for the reactants and products in an **endothermic** reaction.

Complete the diagram by:

   a drawing the reactions profiles for an **uncatalysed** reaction and a **catalysed** reaction

   b labelling the two activation energies and the energy of reaction.

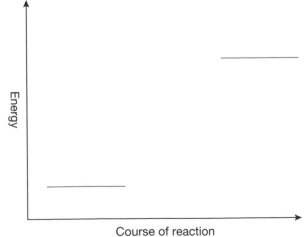

Energy

Course of reaction

Practical: Investigation into how changing the concentration affects the rate of reaction

This is one of the required practicals and you could be questioned on it in the exam. This particular practical places emphasis on you putting forward a hypothesis and making predictions based on it. It also requires the accurate use of appropriate measuring apparatus and the need to make and record a range of measurements using apparatus that measures gas volumes or changes in turbidity. When investigating concentration you should know what factors need to be controlled to make a fair test.

The hypothesis is that increasing the concentration of a reactant in solution increases the rate of reaction. The concentration is your independent variable and is plotted along the horizontal axis (x-axis).

This can be tested by using several different reactions and the practical method requires both the measurement of **gas volume with time** and either a change in **turbidity** or **colour with time**. This is the dependent variable and is plotted up the vertical (y-axis).

SNAPIT!

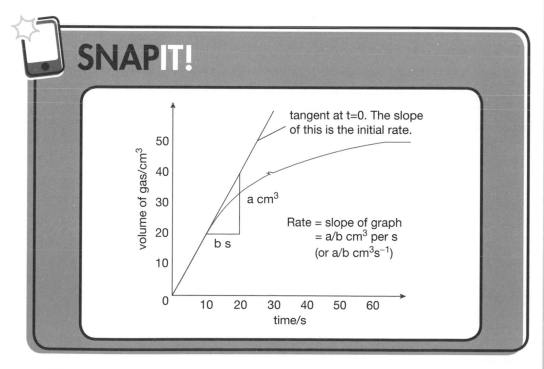

tangent at t=0. The slope of this is the initial rate.

Rate = slope of graph
= a/b cm³ per s
(or a/b cm³s⁻¹)

a cm³

b s

NAILIT!

If the plot of concentration against rate is a straight line then you can say that the **rate is proportional to concentration.**

MATHS SKILLS

You should be able to draw tangents to curves, calculate the slopes of these tangents and express the results in terms cm³/s or g/s.

You must also remember that the independent variable is plotted on the horizontal axis.

Practical Skills

The practical skills that you might be tested on are:

1. The variables you would keep constant to make it a fair test. These are the **control variables.**

   When measuring the effect of concentration other variables must be kept constant. These variables are:

   - the **surface area** and **mass** of any solid reactant
   - the **temperature** of the reactant solution
   - the **acid** used if one of the reactants is an acid
   - the **volume** of acid used.

2. The apparatus you would use to carry out the experiment

   Apparatus used for measuring gas volumes, loss in mass and turbidity is shown in the rates of reaction topic on page 93.

   A **colorimeter** can be used to follow changes in colour in a reaction or if the turbidity changes with time then this change can be followed using a **turbidity meter** or a **light meter**.

3. Measuring the rate.

   The simplest way is to measure the volume of gas given off in a measured time like 1 minute or the loss in mass in 1 minute.

   Another way is to measure the volume of gas with time or mass with time. Plot the results on a graph and then measure the initial rate (see Snap it! for gas volume as an example).

   Measure how long it takes to give a certain volume of gas or change in mass. If this used then the time tells you how slow the reaction is. To give a measure of rate you plot 1/time.

DO IT!

Write short notes or record an MP3 file in which you describe how you would carry out this investigation.

CHECK IT! ✓

1 The reaction between sodium thiosulfate solution and hydrochloric acid produces sulfur as a solid and this turns the solution cloudy. The rate of reaction is measured by the time it takes to obscure a cross on a piece of paper. The results table below shows the results obtained in one experiment.

Volume of $Na_2S_2O_3$ solution/cm³	Volume of hydrochloric acid/cm³	Volume of water/cm³	Time taken to obscure cross/s	Reaction Rate/s⁻¹
5	10	35	128	
10	10	30	64	
20	10		32	
30	10		22	
40	10		16	

Complete the gaps in the table.

What variables are kept constant in this experiment?

Plot a graph of the volume of sodium thiosulfate solution against the rate of reaction and explain what this tells you about the effect of its concentration on the reaction rate. [HINT: The volume of sodium thiosulfate solution is the independent variable.]

Reversible reactions

A reversible reaction is one which can go both ways. This means that as well as reactants forming products, the products can also react to give the reactants.

The reaction where the reactants form the products is called the forward reaction.

The reaction where the products form the reactants is called the reverse reaction.

The symbol used for a reversible reaction is $\rightleftharpoons$.

An example of reversible reactions is the thermal decomposition of ammonium chloride to give ammonia gas and hydrogen chloride gas.

$$NH_4Cl(s) \underset{cool}{\overset{heat}{\rightleftharpoons}} NH_3(g) + HCl(g)$$

If the forward reaction is endothermic then the reverse reaction is exothermic. The opposite is also true.

Example:

$$CuSO_4 5H_2O(s) \underset{exothermic}{\overset{endothermic}{\rightleftharpoons}} CuSO_4(s) + 5H_2O(l)$$

A dynamic equilibrium is when the rate of the forward reaction is equal to the rate of the reverse reaction.

The forward and reverse reactions do not stop they are going on all the time and that is why it is a dynamic equilibrium.

At the same time the concentrations of the reactants and products remain constant.

DOIT!

Write a short account of the experiments shown in the Snap it! box. How can you show that they are reversible reactions?

SNAPIT!

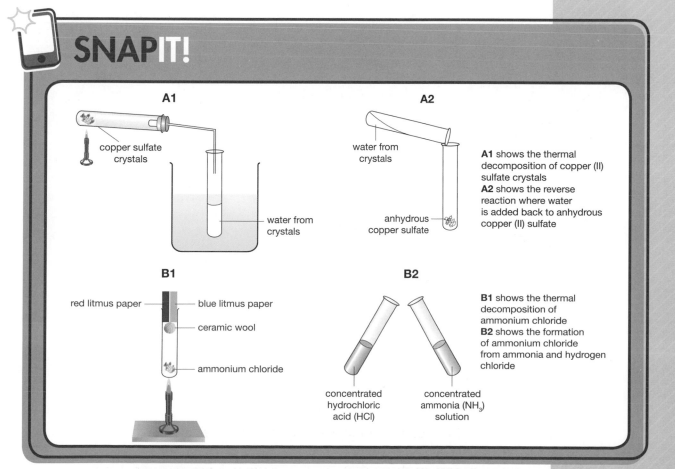

A1
copper sulfate crystals
water from crystals

A2
water from crystals
anhydrous copper sulfate

A1 shows the thermal decomposition of copper (II) sulfate crystals
A2 shows the reverse reaction where water is added back to anhydrous copper (II) sulfate

B1
red litmus paper — blue litmus paper
ceramic wool
ammonium chloride

B2
concentrated hydrochloric acid (HCl)
concentrated ammonia (NH₃) solution

B1 shows the thermal decomposition of ammonium chloride
B2 shows the formation of ammonium chloride from ammonia and hydrogen chloride

CHECK**IT!** ✓

1 What is the symbol that tells you a reaction is reversible?

2 State what is meant by the term reversible reaction.

3 Two compounds A and B react to form Y and Z in a reversible reaction.

   a Complete the equation for the reaction.

     A + B

   b When A is added to B the temperature goes up. What can you say about the forward and reverse reactions?

   c At the beginning when A is added to B why can't there be any reverse reaction?

   d At **equilibrium** which of the following alternatives is/are correct?

     i Both the forward and reverse reactions stop

     ii There are equal amounts of reactants and products

     iii The rate of the forward equals the rate of the reverse reaction

     iv The concentrations of the reactants and products remain constant

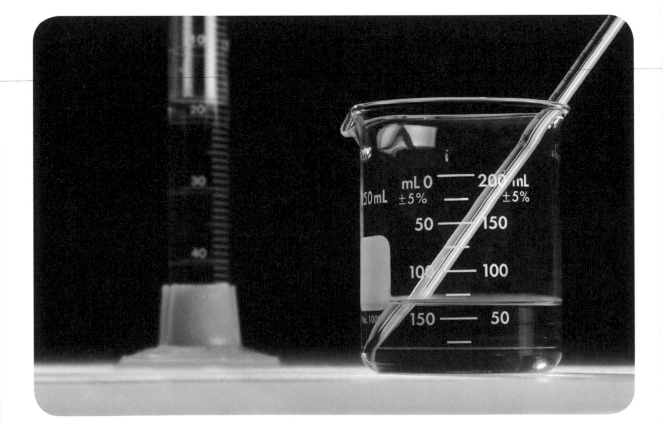

The effect of changing conditions on equilibrium

A chemical system is the reactants and products of a reversible reaction together in a closed container.

If a chemical system is at equilibrium it means that the rate of the forward reaction equals the rate of the reverse reaction.

There are three conditions that can be changed in a chemical system. These are:

• the **temperature** of the system

• the **pressure** of a system if any of the reactants or products are gases

• the **concentration** of any of the reactants or products.

A change in position of equilibrium means that there is a shift towards either the reactants or products.

If a chemical system is at equilibrium and one or more of the three conditions is changed then the position of equilibrium will **shift** so as to cancel out the change and we get either more reactants or more products. This is called Le Chatelier's Principle.

For example:

• If the temperature is increased the equilibrium will shift in favour of the reaction which is accompanied by a decrease in temperature. This means that if the reverse reaction is endothermic then more reactants are formed.

• If the **pressure** is increased the position of equilibrium will shift in favour of the reaction that would lead to a decrease in pressure. This means that the side with **fewer gas molecules** will be favoured. If the numbers of gas molecules on both sides are equal then pressure will have no effect.

• If you increase the **concentration** of one of the products then the system will try and lower its concentration by forming more reactants.

A catalyst has no effect on the position of equilibrium. It speeds up how quickly equilibrium is reached.

A catalyst speeds up both the **forward** and **reverse** reactions **equally**.

DO IT!

Consider the reversible reaction shown below:

$2C_2H_4(g) + O_2(g) \rightarrow 2CH_3CHO(g)$

CH_3CHO is a substance called ethanal. The forward reaction is exothermic.

Explain to someone the best conditions for getting as much ethanal as possible.

NAIL IT!   H

Remember these. If the opposite is done then the reverse will happen.

Increasing the concentration of one of the reactants shifts the equilibrium to the right and favours the forward reaction to make more products.

Increasing the pressure means that the equilibrium will shift to lower the pressure and it does this by making fewer gas molecules. Important – pressure has no effect if the number of gas molecules does not change in the reaction.

Increasing the temperature always favours the endothermic reaction. This is because the endothermic reaction lowers the temperature and this counteracts the change.

SNAPIT!

Consider the industrial manufacture of methanol from carbon monoxide gas (CO) and hydrogen gas (H_2). A copper catalyst is used in this process.

$$CO(g) + 2H_2(g) \xrightleftharpoons[\text{endothermic}]{\text{exothermic}} CH_3OH(g)$$

Consider the effects of some changes on the position of this equilibrium.

H

Change	What happens	Explanation
Increase the concentration of CO gas	Equilibrium shifts to right-hand side $\longrightarrow$	The equilibrium moves in order to lower the concentration of CO gas by reacting it with hydrogen to make more methanol.
Increase the pressure	Equilibrium shifts to right-hand side $\longrightarrow$	If you increase the pressure the system tries to lower it by making fewer gas molecules and this means more methanol is produced.
Decrease the temperature	Equilibrium shifts to right-hand side $\longrightarrow$	If you decrease the temperature the system tries to raise it and this favours the exothermic reaction so making more methanol.

CHECKIT! ✓

H 1 What is a chemical system?

H 2 What does the symbol $\rightleftharpoons$ mean?

H 3 What is meant by the reverse reaction?

H 4 State Le Chatelier's Principle.

H 5 When ethene (C_2H_4) gas reacts with steam (H_2O) at 300°C, ethanol (C_2H_5OH) is formed.

$$C_2H_4(g) + H_2O(g) \xrightleftharpoons[\text{endothermic}]{\text{exothermic}} C_2H_5OH(g)$$

    a What is the effect of increasing the pressure on the amount of ethanol produced? Explain your answer.

    b What is the effect of increasing the temperature on the amount of ethanol produced? Explain your answer.

    c What effect does a catalyst have on the amount of ethanol produced?

REVIEW IT!

1   When calcium carbonate reacts with dilute hydrochloric acid, the gas carbon dioxide is given off as one of the products.

Suggest two ways you could follow the reaction and measure the rate of this reaction.

2   The curve opposite shows the line-graph obtained when a gas is given off during a reaction.

a   How can you calculate the rate of the reaction at any time using the graph?

b   The value of **a** in the graph is 25 cm³ and the value of **b** is 15 s. What is the rate of the reaction at **X**?

Graph obtained when volume of gas is measured with time

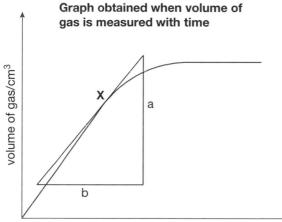

3   The graph opposite shows two identical reactions but at different temperatures.

a   Which of the graphs C or D shows the faster reaction?

b   Explain which graph is the one given at the higher temperature.

c   Using collision theory explain the effect of temperature on reaction rate.

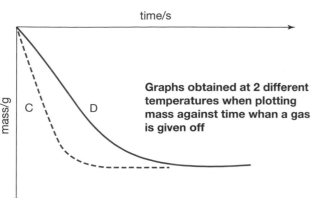

Graphs obtained at 2 different temperatures when plotting mass against time whan a gas is given off

4   When a catalyst is added to hydrogen peroxide solution the following reaction takes place.

$2H_2O_2(aq) \rightarrow 2H_2O(l) + O_2(g)$

a   A catalyst for this reaction is manganese (IV) oxide.

   i   What is meant by the term catalyst?

   ii   If 0.10 g of manganese (IV) oxide is added at the beginning, what is its mass at the end?

b   Draw a reaction profile to show how a catalyst works.

c   Using collision theory explain how a catalyst works

5   You are asked to investigate the effect of concentration on the rate of reaction.

What variables should be kept constant to make it a fair test?

6   When blue copper(II) sulfate crystals (formula = $CuSO_4.5H_2O$) are heated they form water and a white-grey anhydrous copper(II) sulfate powder (formula = $CuSO_4$) is formed.

The equation for the reaction is:

$CuSO_4.5H_2O(l) \rightleftharpoons CuSO_4(s) + 5H_2O(l)$

a   How can you tell that the reaction is reversible?

b   The forward reaction is endothermic. What can you say about the reverse reaction?

c   Give two observations you would make when the water is added back to the white-grey anhydrous copper(II) sulfate.

H 7   When ammonium chloride is heated the following reaction takes place:

$$NH_4Cl(s) \underset{\text{exothermic}}{\overset{\text{endothermic}}{\rightleftharpoons}} NH_3(g) + HCl(g)$$

a   What happens to the position of equilibrium if the temperature is increased?

b   What happens to the position of equilibrium if the pressure is increased?

Organic chemistry

Carbon compounds, hydrocarbons and alkanes

DO IT!

If they are available make ball and stick models (e.g. Molymods) of the first four alkanes using the displayed formulae shown in the Snap it! box.

A **hydrocarbon** is a compound made up of hydrogen and carbon **only**.

Alkanes are hydrocarbons which have the maximum number of hydrogens and no carbon-carbon double bonds. Because of this we call them **saturated** hydrocarbons.

The **general formula** of the alkanes is C_nH_{2n+2}.

The alkanes are a **homologous series**. This means they are a group of compounds with similar **chemical properties**, the same general formula and differ by a CH_2 each time.

They also show a gradation in **physical properties** as the molecules get bigger.

Whatever the homologous series, the way each member of the series is named (prefix) is the same.

The start of each name depends on the number of carbons in the molecule.

Number of carbons in chain	Name starts with
1	Meth–
2	Eth–
3	Prop–
4	But–

Most of the hydrocarbons in crude oil are alkanes.

SNAPIT!

The first four alkanes are listed below along with their molecular formulae and their displayed formulae.

Name	Molecular formula	Displayed formula
methane	CH_4	$H-\overset{\displaystyle H}{\underset{\displaystyle H}{C}}-H$
ethane	C_2H_6	$H-\overset{\displaystyle H}{\underset{\displaystyle H}{C}}-\overset{\displaystyle H}{\underset{\displaystyle H}{C}}-H$
propane	C_3H_8	$H-\overset{\displaystyle H}{\underset{\displaystyle H}{C}}-\overset{\displaystyle H}{\underset{\displaystyle H}{C}}-\overset{\displaystyle H}{\underset{\displaystyle H}{C}}-H$
butane	C_4H_{10}	$H-\overset{\displaystyle H}{\underset{\displaystyle H}{C}}-\overset{\displaystyle H}{\underset{\displaystyle H}{C}}-\overset{\displaystyle H}{\underset{\displaystyle H}{C}}-\overset{\displaystyle H}{\underset{\displaystyle H}{C}}-H$

WORKIT!

What are the molecular and displayed formulae for the alkane with 5 carbons?

The molecular formula can be worked out from the general formula — C_nH_{2n+2}

Here $n = 5$ so $2n+2 = 2 \times 5 + 2 = 12$

Therefore the molecular formula = C_5H_{12}

When you draw the displayed formula the first thing you do is put the 5 carbons in a row.

$$—C—C—C—C—C—$$

Then give each carbon four bonds.

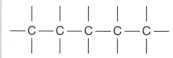

Now add the hydrogen atoms to each end of the bonds.

NAILIT!

The simplest way to get displayed formulae correct is to think that each carbon has got four strong covalent bonds. Also make sure that the lines that represent the covalent bonds go direct from one atom symbol to the other atom symbol.

One of the commonest errors is that candidates write down all the carbons and the bonds coming off from them but forget to draw the hydrogens.

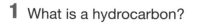

Most of the hydrocarbons in crude oil are alkanes

CHECKIT!

1 What is a hydrocarbon?

2 What do you understand by the term 'homologous series'?

3 State the general formula of the alkanes. What is the molecular formula of the alkane with 6 carbons?

Crude oil, fractionation and petrochemicals

Crude oil is a mixture of many hydrocarbons and this means that it can be separated into its components using a physical method.

These hydrocarbons are miscible and have similar boiling points and therefore they can be separated by fractional distillation.

Most of the hydrocarbons are alkanes.

The separation of crude oil into its constituents is called fractionation and the different parts are called fractions.

Before the crude oil enters the fractionating tower it is heated and evaporates to form a vapour.

As it enters the tower the fractions with the higher boiling points condense lower down the tower to form liquids.

The fractions with lower boiling points continue to rise up the tower until the temperature falls below their boiling point and they also condense.

The fractions that are gases at room temperature leave the tower as gases.

See Snap it! box for details about the fractionation.

All alkanes have simple molecular structures. The boiling points of the fractions depend on the size of the molecules in that fraction. The larger the molecules the greater the intermolecular forces and the higher the boiling points.

Hydrocarbons are good fuels. They undergo complete combustion in air (oxygen) to give carbon dioxide and water as products and this combustion gives out lots of heat energy in an exothermic reaction.

If there is not enough oxygen then incomplete combustion takes place and poisonous carbon monoxide is produced. (This is covered in more detail in the sub topic Atmospheric pollutants).

The apparatus, shown below, is used to test for the products of complete combustion:

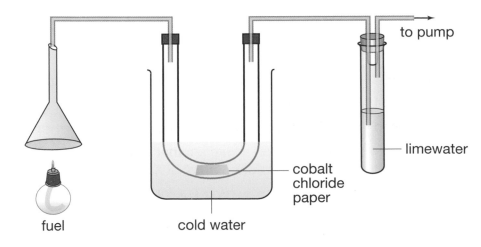

fuel   cold water

to pump

limewater

cobalt chloride paper

WORKIT!

Write out a balanced equation for the combustion of propane (C_2H_3) in oxygen.

The reactants are C_3H_8 and O_2. The products are H_2O and CO_2.

The best way of balancing these equations is to balance the carbons, then the hydrogens and then count up the number of oxygen atoms needed. For the oxygen you can have $\frac{1}{2}O_2$ if odd numbers are needed.

For propane you have 3 carbons and therefore 3 carbon dioxide molecules; you have 8 hydrogens and therefore 4 water molecules. This means that the total number of oxygen atoms is 10 and this means that we need 5 oxygen molecules on the reactant side.

$$C_3H_8(g) + 5O_2(g) \rightarrow 3CO_2(g) + 4H_2O(l)$$

The combustion of ethane (C_2H_6) illustrates the use of $\frac{1}{2}O_2$ because after going through the procedure we find that 7 oxygen atoms are needed on the reactant side.

$$C_2H_6(g) + 3\frac{1}{2}O_2(g) \rightarrow 2CO_2(g) + 3H_2O(l)$$

SNAPIT!

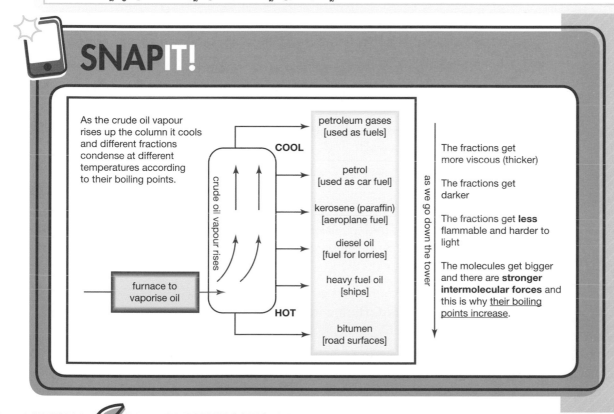

As the crude oil vapour rises up the column it cools and different fractions condense at different temperatures according to their boiling points.

crude oil vapour rises

COOL

petroleum gases [used as fuels]

petrol [used as car fuel]

kerosene (paraffin) [aeroplane fuel]

diesel oil [fuel for lorries]

heavy fuel oil [ships]

HOT

furnace to vaporise oil

bitumen [road surfaces]

as we go down the tower

The fractions get more viscous (thicker)

The fractions get darker

The fractions get **less** flammable and harder to light

The molecules get bigger and there are **stronger intermolecular forces** and this is why their boiling points increase.

CHECKIT!

1 List the uses of the main fractions coming from the fractionating tower.

2 Why is fractional distillation used to separate crude oil into its fractions?

3 What happens to the boiling point of the fractions as we go down the column? Explain why.

4 One fraction X comes off the tower above another fraction Y. Compare the thickness, appearance, ease of lighting and boiling point of X and Y.

5 In the investigation of the products of combustion what are the changes observed in the cobalt chloride paper and the limewater?

6 Complete and balance the following equations for the **complete combustion** of methane (CH_4) and butane (C_4H_{10}).

a $CH_4(g) + \_O_2(g) \rightarrow$ b $C_4H_{10}(g) + \_O_2(g) \rightarrow$

The structural formulae and reactions of alkenes

The homologous series of the alkenes are hydrocarbons with the **functional group C=C**.

Most of the reactions of alkenes depend on the reactions of the C=C group.

Alkenes have the general formula C_nH_{2n}.

They are **unsaturated** hydrocarbons because they have 2 hydrogen atoms less than the maximum number.

Like all hydrocarbons, alkenes burn in air to give carbon dioxide and water. They often burn with a sooty flame because of **incomplete combustion**.

The first four members of the group are shown below:

Name of alkene	Molecular formula	Displayed structural formula
ethene	C_2H_4	
propene	C_3H_6	
butene	C_4H_8	
pentene	C_5H_{10}	

DO IT!

Use Molymods or ball and stick models to show these reactions so you can visualise what is happening.

The main type of reaction of alkenes is **addition**.

In addition reactions the double C=C bond becomes a single C–C bond and the other reactant splits into two. One part of the reactant bonds to one carbon of the C=C bond and the other part bonds to the other carbon.

The test for alkenes is to add **bromine water**. Alkenes **decolourise** the bromine water, which means the bromine water goes from orange to colourless. This is an addition reaction.

SNAP IT!

alkene → reactant

Examples of X–Y.

In steam (H_2O) X = H and Y = OH

In Br_2 X = Br and Y = Br

NAILIT!

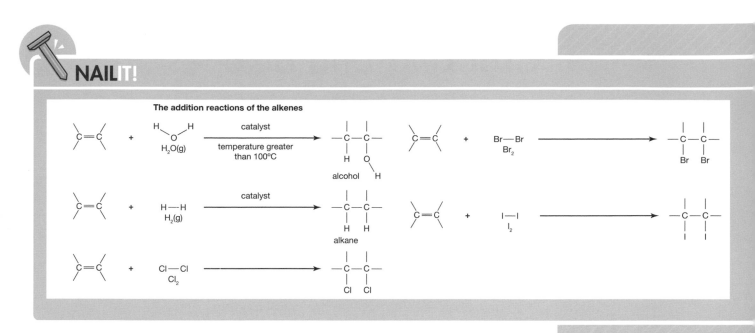

The addition reactions of the alkenes

WORKIT!

Complete the following reactions using structural formulae. Think of the molecule being added as X—Y. X adds to one carbon and Y to the other.

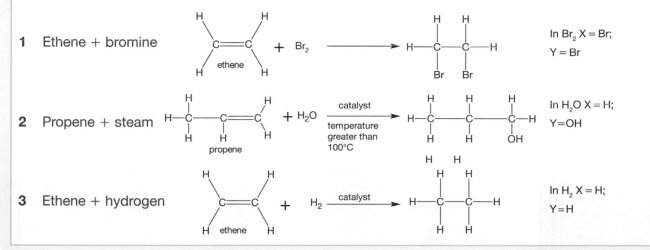

1 Ethene + bromine

In Br_2 X = Br; Y = Br

2 Propene + steam

In H_2O X = H; Y = OH

3 Ethene + hydrogen

In H_2 X = H; Y = H

✓ CHECKIT!

1 State the general formula of the alkenes.

2 a Name the alkene with 4 carbons.

  b Explain why methene does not exist.

3 a Why are alkenes unsaturated hydrocarbons?

  b i Give the formula of the alkene with 4 carbons.

    ii When the alkene with 4 carbons burns in insufficient oxygen it undergoes **incomplete combustion** to give carbon and water as the products. Write the balanced symbol equation for the reaction.

4 Name the main type of reaction of alkenes.

5 a Draw the structural formula of the product of the reaction of ethene with steam.

  b Write the balanced symbol equation for the reaction.

Cracking and alkenes

Cracking is the breaking down of large **alkane** molecules into **smaller** alkanes and **alkenes**.

The smaller alkanes make good fuels and the alkenes are used to make polymers.

Cracking is an important process for two main reasons:

1 **It converts fractions which have a low demand into higher demand fractions.** For example, after fractional distillation there is not enough of the petrol fraction for use as a fuel but there is more than required of the kerosene fraction. This means that some of the kerosene can be cracked to give alkanes that make up the petrol fraction.

2 **It makes useful hydrocarbons not naturally found in crude oil.** For example, cracking also gives alkenes which are not found in crude oil but are very important in the manufacture of polymers.

The general formula of alkenes is C_nH_{2n}. Remember the general formula for alkanes is C_nH_{2n+2} and the two fewer hydrogen atoms is because of the C=C double bond in the alkene molecule.

Cracking improves the economic value of the fractions coming off the tower.

In cracking the alkane vapours are passed over a heated catalyst or mixed with steam and heated up to a high temperature.

DOIT!

Refer to your notes and textbook to analyse the results of the experiment in the Snap it! box.

SNAPIT!

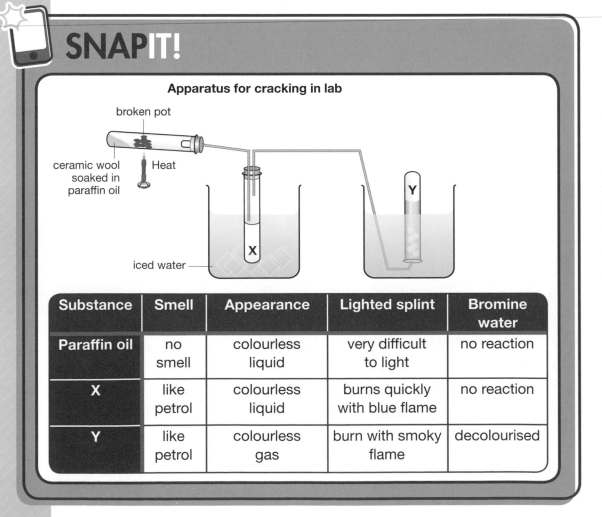

Apparatus for cracking in lab

Substance	Smell	Appearance	Lighted splint	Bromine water
Paraffin oil	no smell	colourless liquid	very difficult to light	no reaction
X	like petrol	colourless liquid	burns quickly with blue flame	no reaction
Y	like petrol	colourless gas	burn with smoky flame	decolourised

MATHS SKILLS

Balancing equations: You may have to use general formulae to identify the products in a cracking reaction.

Remember that the general formula of the alkanes is C_nH_{2n+2} and for the alkenes it is C_nH_{2n}.

WORKIT!

In the following reactions identify the alkanes and alkenes:

$$C_8H_{18} \rightarrow C_4H_8 + C_4H_{10}$$

Alkane Alkene   Alkane

$$C_{10}H_{22} \rightarrow C_2H_4 + C_4H_8 + C_4H_{10}$$

Alkane   Alkene Alkene Alkane

An oil refinery

NAILIT!

The important thing about cracking is that it turns chemicals that are not very useful into ones that are very useful and economically important.

The amounts of short chain alkanes which are produced by fractionation of crude oil are not enough to meet demands. Cracking overcomes this shortfall.

✓ CHECKIT!

1 Define cracking.

2 Describe how cracking is carried out in industry.

3 a What is the general formula for alkenes?

  b Describe the chemical test for alkenes.

  c Give a use for alkenes.

4 Describe the economic importance of cracking.

5 Complete the following equations and identify the alkanes and alkenes in the reactions.

  a $C_6H_{14} \rightarrow C_2H_6 + \underline{\hspace{2cm}}$

  b $C_8H_{18} \rightarrow C_2H_4 + C_3H_6 + \underline{\hspace{2cm}}$

  c $C_{12}H_{26} \rightarrow 3C_2H_4 + \underline{\hspace{2cm}}$

Alcohols

The **alcohols** are a homologous series and their functional group is the –OH group.

This means that they have similar chemical properties.

As the number of carbons increase there is a gradual change in physical properties. For example, their boiling points get higher.

Their general formula is $C_nH_{2n+2}O$.

Alcohols have lots of uses as fuels and **solvents** and **ethanol** is used in alcoholic drinks.

Ethanol is the most important alcohol and it can be produced by the **fermentation** of aqueous sugar solution using **yeast**.

The process works best at around 40°C. When the ethanol concentration gets to about 15% the yeast is killed and the reaction stops. Oxygen should be excluded from the apparatus because the fermentation process is **anaerobic**.

mixture of yeast and sugar solution

limewater

Combustion – alcohols burn in air to give carbon dioxide and water:

e.g. $CH_3CH_2OH(l) + 3O_2(l) \rightarrow 2CO_2(g) + 3H_2O(l)$

The alcohols tend to burn with a non-luminous flame.

Alcohols with less than four carbons are very soluble in water and dissolve to give colourless solutions.

Alcohols can be oxidised to give **carboxylic acids**. The carboxylic acid formed has the same number of carbons as the alcohol.

If the **oxidising agent** is purple acidified potassium manganate(VII) then it is decolourised. If it is acidified potassium dichromate(VII) then it changes from orange to green.

When **sodium** is added to alcohols there is (effervescence) fizzing because hydrogen gas is produced and the sodium disappears to leave a white solid.

DO IT!

Look up your notes on fermentation. You might have biology notes on the topic as well. Use the diagram on this page to help you.

OR

Make sketches of what you see when alcohols react with sodium, burn in air and are oxidised by potassium manganate(VII) or potassium dichromate(VI).

SNAPIT!

Alcohol molecules can be represented in two ways.

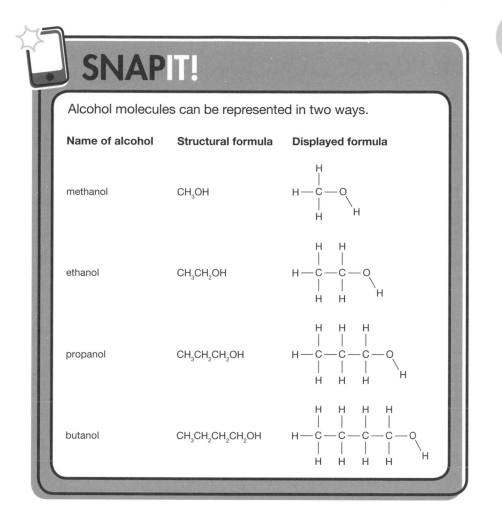

Name of alcohol	Structural formula	Displayed formula
methanol	CH_3OH	
ethanol	CH_3CH_2OH	
propanol	$CH_3CH_2CH_2OH$	
butanol	$CH_3CH_2CH_2CH_2OH$	

CHECKIT!

1 Name the alcohol with 3 carbons.

2 What is the functional group in alcohols?

3 Draw the structural and displayed formulae for the alcohol which has the molecular formula C_2H_6O.

4 Write the balanced symbol equation for the combustion of:

   a $CH_3CH_2CH_2CH_2OH$

   b CH_3OH.

5 Draw and explain the apparatus you would use in the laboratory to carry out the fermentation of glucose.

6 a If ethanol is oxidised what carboxylic acid is formed?

   b What would you see happen if the oxidising agent used is:

   i acidified potassium manganate(VII)

   ii acidified potassium dichromate(VI)?

Carboxylic acids

The carboxylic acids are a homologous series with the functional group $-COOH$.

Their general formula is $C_nH_{2n}O_2$

An acid is a substance that produces hydrogen ions when dissolved in water.

The carboxylic acid ($-COOH$) group gives hydrogen ions when the acid dissolves in water.

But in aqueous solution only a small proportion of carboxylic acid molecules ionise to give hydrogen ions.

For this reason they are weak acids and in reactions they react more slowly than strong acids such as hydrochloric acid.

Solutions of carboxylic acids always have higher pH values than those of strong acids that have the same concentration.

As acids they react with carbonates to form a salt, carbon dioxide and water. Because carbon dioxide gas is produced this reaction gives effervescence (fizzing).

Carboxylic acids react with alcohols to produce esters. For example, ethanol and ethanoic acid react to form ethyl ethanoate and water. This reaction needs a strong acid as a catalyst.

If the carboxylic acid is heated with an alcohol and a strong acid catalyst like sulphuric acid, an ester is formed and can be detected by its fruity smell.

SNAPIT!

Name of carboxylic acid	Structural formula	Displayed formula
methanoic acid	HCOOH	
ethanoic acid	CH_3COOH	
propanoic acid	CH_3CH_2COOH	
butanoic acid	$CH_3CH_2CH_2COOH$	

NAILIT!

It is the hydrogen ions that are responsible for the reactions of acids, so carboxylic acids will react more slowly than strong acids like hydrochloric acid.

STRETCHIT!

Carboxylic acids are weak acids because they are only partially ionised in water. For example, in a solution of ethanoic acid only 4 out of every 1000 molecules produce hydrogen ions. A solution of the same concentration of hydrochloric acid, which is a strong acid, is fully ionised.

CHECKIT!

1 What is the functional group of carboxylic acids?

2 Draw the structural and displayed formulae for the carboxylic acid with 2 carbons.

3 When calcium carbonate is added to an aqueous solution of ethanoic acid there is **fizzing** and **the reaction is slow.** Explain these observations.

Addition polymerisation

In addition polymerisation small alkene molecules called monomers join together to make very large molecules called polymers.

In addition polymerisation there is only one product.

In an addition polymer there is a repeating unit and this has the same atoms as the monomer.

A polymer is named by simply placing 'poly' in front of the name of the monomer. For example, the polymer formed from propene is called **polypropene**.

SNAPIT!

The general equation for addition polymerisation is shown below. W, X, Y and Z can be any non-metal atom or group of atoms:

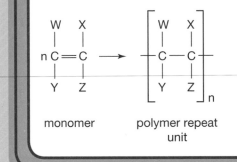

monomer

polymer repeat unit

FOUR important things about the formula of the polymer repeat unit:
1. The n shows there are many repeat units
2. The atoms or groups on the double bond are in the correct order
3. The double bond is now a single bond
4. <u>Do not forget</u> the bonds on each side that show the unit is attached to others on both sides

NAILIT!

In the product, W, X, Y and Z are unchanged and in the same positions. The C=C has become C—C and the spare bonds at each end show that the unit carries on bonding to other units on either side. The lower case 'n' is a large number.

If you are asked to draw the alkene monomer from the repeat unit, the reverse procedure applies. The C—C becomes C=C and the W, X, Y and Z stay in the same positions. Do not forget to remove the bonds at each end.

Plastics are examples of polymers

WORKIT!

Show the repeat units for the polymers formed from the following monomers:

a) ethene   b) chloroethene

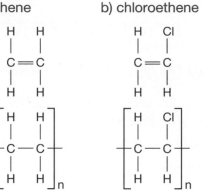

Draw the monomers used to make the following polymers:

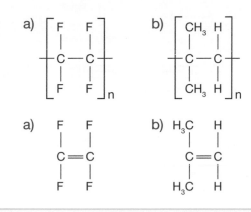

✓ CHECKIT!

1 What is a monomer?

2 Draw the repeating unit for the polymer formed from the alkene dichlorethene (see below).

3 Draw the alkene responsible for making the following polymer (see below).

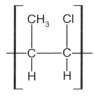

4 Name the polymer formed from butene.

Condensation polymerisation

H

A **condensation** reaction takes place when two molecules react to form a larger molecule and a small molecule such as water.

An example is the reaction between an alcohol and a carboxylic acid to form an ester and water.

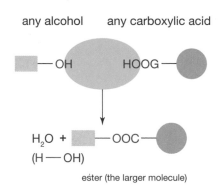

If the alcohol has two —OH groups on it and the carboxylic acid has two —COOH groups on it then the condensation reaction can carry on and on to form a polyester.

The general polymers HO —⬜— OH and HOOC —⬤— COOH react as shown below

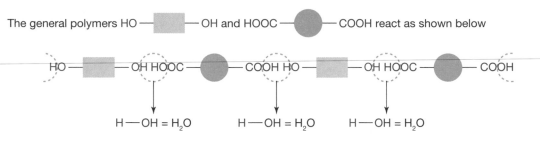

Each alcohol bonds to 2 carboxylic acids and each carboxylic acid bonds to 2 alcohols. This happens many many times to form a condensation polymer

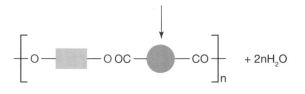

A **polyester** is a **condensation polymer**.

The alcohol and the carboxylic acid molecules are the monomers.

Examples of monomers for condensation polymerisation are $HOCH_2CH_2OH$ and $HOOCCH_2CH_2CH_2CH_2COOH$.

DO IT!

Write a short description of the main features of addition polymerisation and condensation polymerisation.

Nylon rope is an example of a polymer formed by condensation polymerisation

SNAPIT!

The formation of a condensation polymer such as a polyester is shown below:

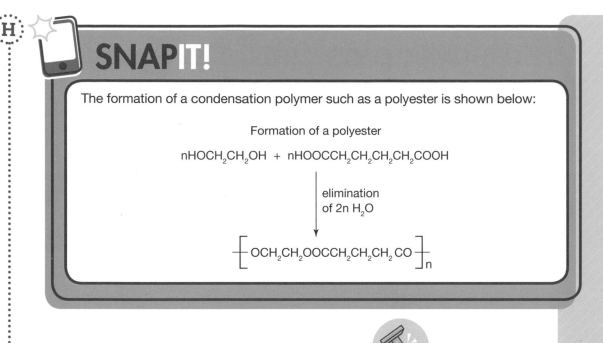

Formation of a polyester

$$nHOCH_2CH_2OH + nHOOCCH_2CH_2CH_2CH_2COOH$$

elimination
of $2n$ H_2O

$$-[OCH_2CH_2OOCCH_2CH_2CH_2CO]_n-$$

WORKIT!

If the alcohol is $HOCH_2CH_2OH$ and the carboxylic acid is $HOOCCH_2CH_2COOH$ then the structure of the condensation polymer is:

$$-[OCH_2CH_2OOCCH_2CH_2CO]_n-$$

What are the 2 monomers that would form the polymer

$$-[OCH_2CH_2CH_2OOCCH_2CH_2CH_2CH_2CO]_n-\ ?$$

The 2 monomers are $HOCH_2CH_2CH_2OH$ (the alcohol) and $HOOCCH_2CH_2CH_2CH_2COOH$ (carboxylic acid).

NAILIT!

The link between two monomers in a polyester is the ester link or –OOC– group of atoms. So if you have two monomers, an alcohol and a carboxylic acid, then the OOC links together the alcohol and the carboxylic acid. As each ester link is formed one water molecule is eliminated. If n polymer repeat units are formed then 2n links are formed and 2n water molecules are eliminated. The final condensation polymer has the following general structure:

$$-[O\text{-alcohol chain-}OOC\text{-carboxylic acid chain-}CO]_n-$$

Reversing the procedure you can examine the structure of the polymer and draw the structures of the monomers HO–alcohol chain–OH and HOOC–carboxylic acid chain–COOH

CHECKIT!

H 1 **a** **i** What is a condensation reaction?

   ii Name the simple molecule eliminated when a polyester is formed.

b Give the main differences between addition polymerisation and condensation polymerisation.

H 2 What is the condensation polymer formed from the two monomers shown below?

HOOC-COOH and $HOCH_2CH_2OH$

H 3 Draw the monomers responsible for forming the polymer shown below.

$$-[OCH_2OOCCO]_n-$$

Amino acids and DNA

Amino acids have at least two functional groups in their molecules.

Amino acids are the monomers that form condensation polymers called **polypeptides** or **proteins**. The small molecule eliminated each time is water.

There are 20 different naturally occurring amino acids and these can be linked in different sequences giving different proteins.

DNA gives genetic instructions from which the essential proteins in our body are made.

DNA is a large molecule which consists of two polymer chains which together form a **double helix**.

Each DNA chain is a condensation polymer of four different monomers called **nucleotides**. All the large molecules found in organisms are condensation polymers. Examples are starch and cellulose which are polymers of the monomer glucose.

DO IT!

Look at your biology notes about DNA and proteins, and add any additional notes that help your understanding of these polymers.

SNAP IT!

The amino acid glycine H_2NCH_2COOH forms the condensation polymer $-(NHCH_2CO)_n-$

A protein contains different amino acids in a condensation polymer. The amino acids are linked by the peptide bond:

$-NHCO-$

CHECK IT!

1 In organisms what small molecule is always eliminated in condensation polymerisation?

2 Polysaccharides are polymers of sugars. Examples are starch and cellulose. The diagram below is a simplified version of monomer sugar rings. Draw the condensation polymer formed from these monomers.

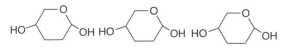

3 Give a brief description of DNA as a polymer.

4 Give the repeating unit found in polyglycine.

Organic chemistry

1 a Determine the molecular formulae of the compounds A to D shown below.

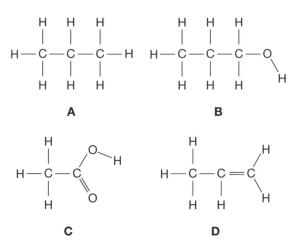

A   **B**

C   **D**

b Which of the compounds A to D is or are:

i a hydrocarbon

ii an alkene

iii a carboxylic acid

iv an alcohol

v an alkane?

c Name all four compounds.

2 a Identify the functional group in an alkene.

b Why are the alkenes considered a homologous series?

c Describe the chemical test for alkenes.

d Complete the following reactions:

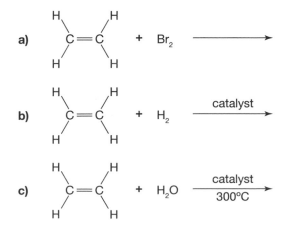

3 a State the physical property which allows us to separate crude oil into its fractions using fractional distillation.

b As we go up the fractionating tower what happens to the following properties:

i viscosity

ii boiling point

iii ease of lighting?

c Explain the trend in boiling point as you go up the tower.

4 Alkanes are the main components of crude oil.

a What is the general formula of the alkanes?

b Complete and balance the following equations for the complete combustion of the two hydrocarbons methane and ethane:

i $CH_4(g) + O_2(g) \rightarrow$

ii $C_2H_6(l) + O_2(g) \rightarrow$

c The diagram below shows the apparatus used to investigate what is formed when an alkane burns in air.

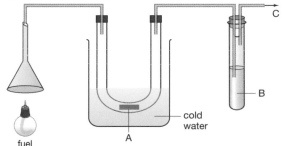

i Give the correct labels for A, B and C.

ii Describe what happens to A. What does the change show?

iii Describe what happens to the liquid in B. What does the change show?

5 a Why is it necessary to carry out cracking on alkanes which have large molecules?

b Complete the following equations:

i $C_{10}H_{22} \rightarrow C_4H_8 +$

ii $\rightarrow C_3H_6 + C_6H_{14}$

Chemical analysis

Pure substances and formulations

DO IT!

Look up the ingredients of a simple over-the-counter drug. A painkiller like Nurofen is a good one to do. How do the ingredients of the formulation improve the drug? Google 'The ingredients of Nurofen'.

Pure substances are either single elements or single compounds.

Everyday descriptions of pure substances are inaccurate. For example, pure spring water is a solution of various minerals and gases.

Pure substances melt and boil at specific temperatures.

Mixtures melt and boil over a range of temperatures.

A formulation is a mixture that is designed as an improvement upon a pure substance on its own.

Examples of formulations are metal alloys, drugs and paints.

For example, a drug has in its formulation the active chemical and other substances that stop the drug from going off and make it easy to swallow.

MATHS SKILLS

Using formulae and inserting values into them using percentages.

The formulae used are:

$$\text{Percentage} = \frac{\text{mass of component}}{\text{total mass}} \times 100\%$$

$$\text{Number of moles} = \frac{\text{mass}}{Mr}$$

WORKIT!

A drug formulation in tablet form weighs 500 mg. The two components are 350 mg of a stabiliser (X) with a relative formula mass 70, and the 150 mg of the active drug itself (Y) which has a relative formula mass of 125.

What is the percentage composition of the tablet in terms of:

a mass

The percentage by mass of X = 350/500 = 70% = 70%

Therefore the percentage by mass of Y = 100 - 70 = 30%

b moles? **H**

The number of moles of X = mass/M_r = 350 × 10⁻³/70 = 5 × 10⁻³ mol

The number of moles of Y = mass/M_r = 150 × 0⁻³/125 = 1.2 × 10⁻³ mol

Therefore the total number of moles = 5 × 10⁻³ mol + 1.2 × 10⁻³ mol = 6.2 × 10⁻³ mol

Mole percentage for X = (5 × 10⁻³/6.2 × 10⁻³) × 100% = 80.6%

This means that the mole percentage for Y = 100 - 80.6% = 19.4%

SNAPIT!

A distinctive melting point is a criterion for purity. The apparatus used is shown below along with a typical heating curve for a **pure** substance.

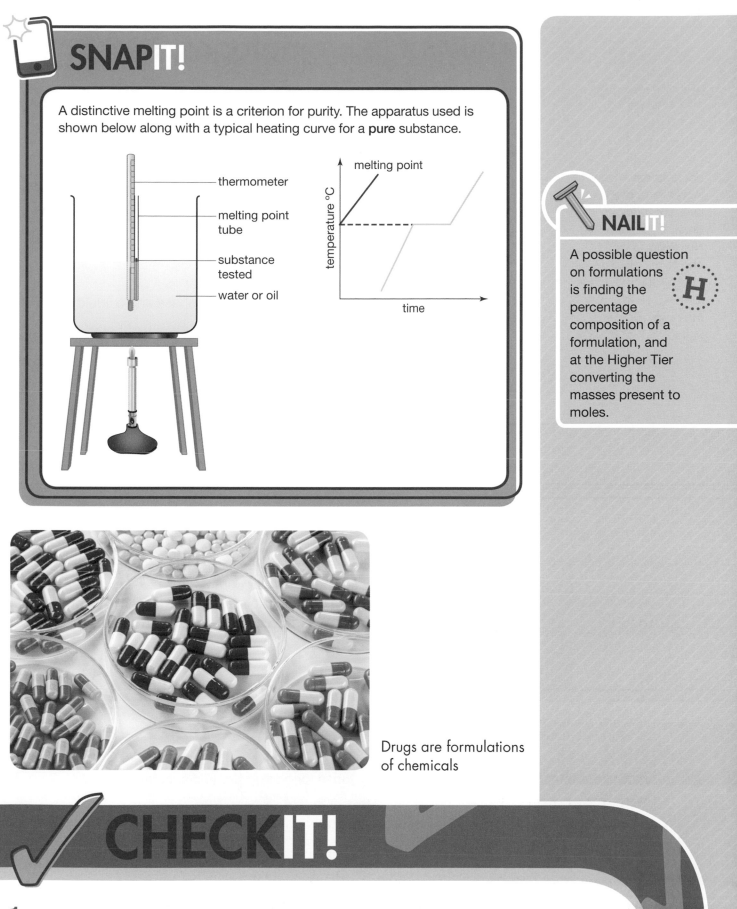

- thermometer
- melting point tube
- substance tested
- water or oil

melting point

temperature °C

time

NAILIT!

A possible question on formulations is finding the percentage composition of a formulation, and at the Higher Tier converting the masses present to moles.

H

Drugs are formulations of chemicals

CHECKIT!

1 What is a pure substance?

2 Why isn't 'pure orange juice' really pure?

3 What is a formulation?

4 Give one example of a product which is a formulation and give brief details of its composition.

Practical: Chromatography

This is one of the required practicals and you could be questioned on it in the exam. The practical emphasises that you carry out a practical safely and accurately, make measurements and make conclusions using these measurements.

Chromatography is a technique that can be used to separate mixtures into their components and identify these components.

Chromatography involves two phases – a stationary phase and a mobile phase.

In paper chromatography, paper is the stationary phase and a liquid solvent is the mobile phase.

When substances are added to the paper the mobile phase carries them through the paper. The distance a compound moves on the paper depends on its relative attraction for the paper and the solvent. The R_f value for a substance is equal to the distance moved by its spot divided by the distance moved by the solvent.

Compounds that have a higher attraction for the paper and a low attraction for the solvent spend a lot of time on the paper and move up the paper slowly.

Compounds that have a higher attraction for the solvent and have less of an attraction for the paper move quickly up the paper.

At the end of the experiment the spots obtained after the solvent is run up the paper is called a chromatogram.

Mixtures give more than one spot and pure compounds give only one spot. If the substances being separated are colourless then a locating agent is needed to show how far they have moved. Sometimes a UV lamp can be used.

Practical Skills

Paper chromatography is often used to separate a mixture of coloured substances such as those found in inks and food colourings.

Chromatography requires a container, a supply of solvent(s), a ruler for measuring R_f values, chromatography paper and very thin capillary tubes for adding the test substances to the paper.

A pencil line is drawn a short distance from the bottom of the paper and this is called the baseline. The substances to be investigated are added at regular intervals along this line.

If the starting spots are too large then it will be very difficult to separate the mixture into easily identified substances. This is because the spots get larger as they rise up the paper.

The choice of solvent is not limited to water and should be chosen on the basis that it gives a good separation.

When measuring the R_f value of a substance the measurement is taken from the line where the substances are added to the paper and the middle of the spot obtained after the solvent has run up the paper.

SNAPIT!

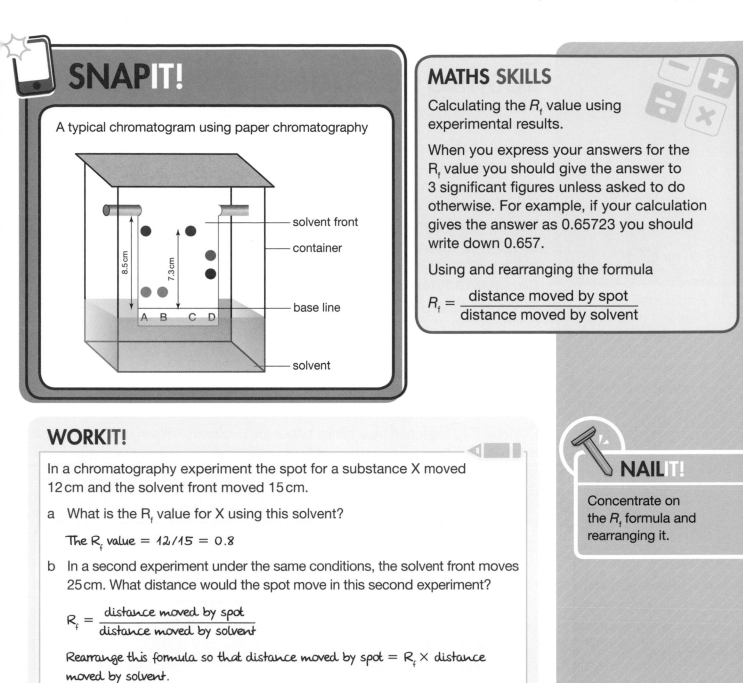

A typical chromatogram using paper chromatography

solvent front
container
base line
solvent

8.5 cm   7.3 cm

A B   C D

MATHS SKILLS

Calculating the R_f value using experimental results.

When you express your answers for the R_f value you should give the answer to 3 significant figures unless asked to do otherwise. For example, if your calculation gives the answer as 0.65723 you should write down 0.657.

Using and rearranging the formula

$$R_f = \frac{\text{distance moved by spot}}{\text{distance moved by solvent}}$$

WORKIT!

In a chromatography experiment the spot for a substance X moved 12 cm and the solvent front moved 15 cm.

a What is the R_f value for X using this solvent?

The R_f value = 12/15 = 0.8

b In a second experiment under the same conditions, the solvent front moves 25 cm. What distance would the spot move in this second experiment?

$$R_f = \frac{\text{distance moved by spot}}{\text{distance moved by solvent}}$$

Rearrange this formula so that distance moved by spot = R_f × distance moved by solvent.

This means that the distance moved by spot = 0.8 × 25 cm = 20 cm.

NAILIT!

Concentrate on the R_f formula and rearranging it.

CHECKIT!

1 In paper chromatography what is the stationary phase and what is the mobile phase?

2 The base line is drawn in pencil. Why do you not draw the baseline in pen?

3 This question concerns the chromatogram shown in the Snap it! box.

a Which of the three substances B, C and D is/are a pure substance? Explain your answer.

b i How can you tell that A is a mixture?

ii What substances make up the mixture A?

c Calculate the R_f value for substance C.

Testing for gases

Several chemical reactions produce a gas or gases. In solutions this produces effervescence (fizzing).

Hydrogen is a flammable gas. The test for hydrogen is a burning splint which is extinguished (put out) with a 'pop'.

The equation for this combustion reaction is: $2H_2(g) + O_2(g) \rightarrow 2H_2O(l)$

Oxygen (O_2) supports combustion. The test for oxygen is a glowing splint which relights in oxygen.

Carbon dioxide reacts with a solution of limewater to give solid calcium carbonate. When carbon dioxide is passed into limewater the limewater turns cloudy or milky.

The equation for this reaction is:

$CO_2(g) + Ca(OH)_2(aq) \rightarrow CaCO_3(s) + H_2O(l)$

                colourless   white solid

                   liquid

Chlorine (Cl_2) bleaches blue litmus paper (or UI paper). This is used as the test for chlorine.

DO IT!

Print out the diagrams of the tests in the Snap it! box and then write down a list of comments to accompany the m.

SNAP IT!

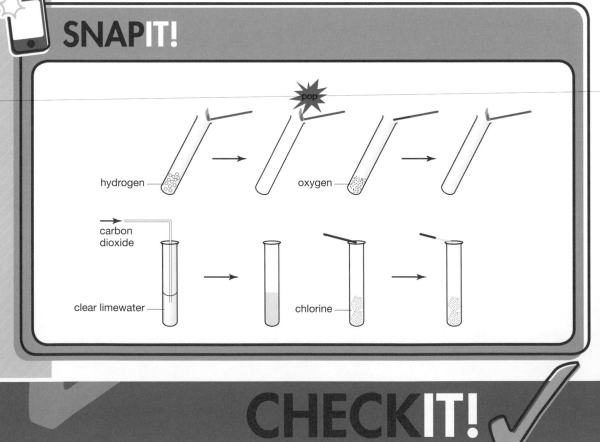

CHECK IT!

Describe how you would test for the gases produced in the following reactions:

a $2H_2O_2(aq) \rightarrow 2H_2O(l) + O_2(g)$

b $CaCO_3(s) + 2HCl(aq) \rightarrow CaCl_2(aq) + CO_2(g) + H_2O(l)$

c $Mg(s) + 2HCl(aq) \rightarrow MgCl_2(aq) + H_2(g)$

d $2NaCl(aq) + 2H_2O(l) \rightarrow 2NaOH(aq) + H_2(g) + Cl_2(g)$

Identifying metal ions using flame tests and flame emission spectroscopy

Positive metal ions are often referred to as cations because they move towards the cathode (the negative electrode) in electrolysis.

When the salts of some metals are heated in a hot Bunsen burner flame they give a distinctive characteristic colour that can be used to identify the metal ion present. This is called a flame test.

When carrying out the flame test a very hot (roaring blue) Bunsen flame must be used.

If the flame test is used to identify the metals in a mixture some flame colours are masked by others. For example, the lilac flame given by potassium ions is easily masked by the stronger yellow colour of sodium ions.

DOIT!

Devise a method or make a short MP3 recording of how you would do a flame test.

SNAPIT!

Metal ion	Colour of flame	Approximate colour
Li^+	crimson	
Na^+	yellow	
K^+	lilac	
Ca^{2+}	orange-red	
Cu^{2+}	green	

NAILIT!

Learn the colours of the flames. There could well be a question on identifying salts and the flame colours would form part of the answer.

Flame emission spectroscopy uses the same principles as the flame test but uses instrumentation to do it.

A small sample is put into a flame and the light given out is passed through a spectroscope. A series of coloured lines called a line spectrum is obtained and one of these lines is the one that is used to analyse the sample.

The intensity of the line gives a measure of the concentration of the ion.

The advantages of flame emission spectroscopy over the simple flame test are:

- It is more sensitive.

- It can be used to measure the concentration of the ion in the sample.

- It can look at distinctive areas of the colour spectrum emitted by a heated element and not by others. This overcomes the problem of some colours being masked by others.

- This means that it can be used to analyse the composition of mixtures which cannot be done using the flame test.

MATHS SKILLS

One possible question about instrumentation could involve figures for a calibration graph and using them to find an unknown concentration.

WORKIT!

On the graph paper provided draw a calibration line for the following readings from a flame emission spectroscope.

Conc.mg/cm³	0	0.10	0.20	0.40	080	1.00
Line intensity	0	0.026	0.051	0.103	0.207	0.25

A solution X gave a reading of 0.078. What is its concentration?

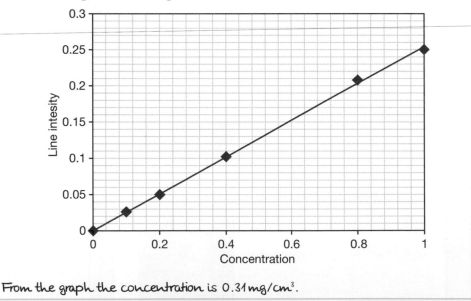

From the graph the concentration is 0.31 mg/cm³.

CHECKIT! ✓

1 State the colours of Ca^{2+} and K^+ ions in the flame test.

2 List some advantages of instrumentation when detecting ions in the flame test.

Identifying metal ions using sodium hydroxide solution

When aqueous sodium hydroxide solution is added to the solutions of some metals, precipitates are formed and the colours of these precipitates can be used to identify the metal ion present.

If the formula of the metal ion is M^{n+} then the formula of the precipitate is $M(OH)_n$.

The colours of these precipitates are summarised below:

Name and formula of ion	Colour of precipitate	Extra comments
Magnesium Mg^{2+}	White	Same as calcium but gives no colour in flame test
Calcium Ca^{2+}	White	Gives an orange-red colour in flame test
Aluminium Al^{3+}	White	The white precipitate dissolves when excess (extra) sodium hydroxide solution is added
Copper(II) Cu^{2+}	Blue	
Iron(II) Fe^{2+}	Green	
Iron(III) Fe^{3+}	Brown	

SNAPIT!

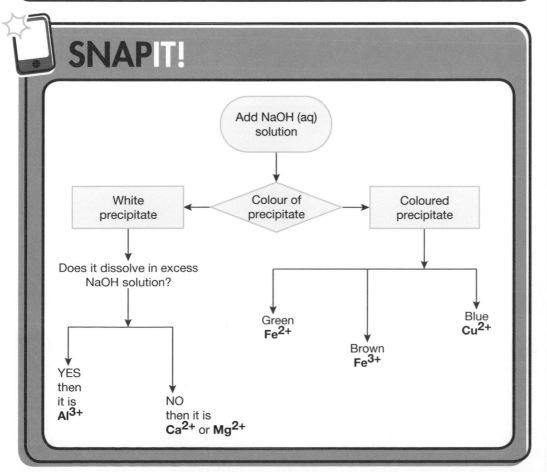

Add NaOH (aq) solution

Colour of precipitate

White precipitate

Coloured precipitate

Does it dissolve in excess NaOH solution?

Green Fe^{2+}

Brown Fe^{3+}

Blue Cu^{2+}

YES then it is Al^{3+}

NO then it is Ca^{2+} or Mg^{2+}

DOIT!

The flowchart in the Snap it! box does not have to be the one you use. See if you can make up your own.

Copy or print out the flowchart in the Snap it! box. Add colour to it. Also add in the other cations that give coloured flames in the flame test.

STRETCH IT!

If you are asked to write the ionic equation for a precipitation reaction then only the metal ion and hydroxide ions are involved. The anion in the metal salt and the sodium ions in the sodium hydroxide are spectator ions because they do not take part in the reaction.

Ionic equations were covered in displacement reactions of the halogens and in the reactivity series.

For example:

$Mg^{2+}(aq) + 2OH^-(aq) \rightarrow Mg(OH)_2(s)$

$Fe^{3+}(aq) + 3OH^-(aq) \rightarrow Fe(OH)_3(s)$

NAIL IT!

When you write the symbol equations for these reactions you should know that the metal and the hydroxide are always listed together. State symbols are also important here. The precipitate which is the hydroxide is the solid (s) and all the other substances are in solution (aq).

For example:

$2NaOH(aq) + MgSO_4(aq) \rightarrow Mg(OH)_2(s) + Na_2SO_4(aq)$

A green precipitate settling in solution

CHECK IT!

1 Describe the colour of the precipitate you would get if you add sodium hydroxide solution to the following solutions:

   a $CaCl_2(aq)$ b Iron(III) nitrate

   c $CuSO_4(aq)$

2 How can you distinguish between two solutions which contain magnesium and aluminium ions?

3 Give the word and balanced symbol equations for the reaction of copper(II) sulfate ($CuSO_4$) solution with sodium hydroxide (NaOH) solution.

H 4 Give the ionic equation for the reaction of sodium hydroxide solution with $CuSO_4$ solution.

H 5 When a solution containing green Fe^{2+} ions is left overnight the solution gradually turns brown in colour.

   a What is contained in the new solution?

   b Explain why this is an oxidation reaction.

Testing for negative ions (anions) in salts

Negative ions can also be called anions because they move towards the anode (positive electrode) in electrolysis.

The five ions that are tested for are the three halide ions (Cl^-, Br^- and I^-), sulfate ions (SO_4^{2-}) and carbonate ions (CO_3^{2-}).

The tests for the halide ions and sulfate ions require you to add an acid before the testing solution. This is to remove any carbonate ions that would also react with the testing solution.

The acid used contains the same negative ion as the testing solution. This means that nitric acid is added before silver nitrate and hydrochloric acid is added before barium chloride solution.

The tests are shown below:

Ion tested for	Testing reagents	Results of positive test
Chloride (Cl^-)	Add nitric acid followed by silver nitrate solution	White precipitate (of silver chloride)
Bromide (Br^-)	Add nitric acid followed by silver nitrate solution	Cream precipitate (of silver bromide)
Iodide (I^-)	Add nitric acid followed by silver nitrate solution	Yellow precipitate (of silver iodide)
Sulfate (SO_4^{2-})	Add hydrochloric acid followed by barium chloride solution	White precipitate (of barium sulfate)
Carbonate (CO_3^{2-})	Add hydrochloric acid then pass gas formed through limewater	Effervescence (fizzing) and gas produced turns limewater cloudy/milky.

DO IT!

It's a good idea to review previous topics before you revise this one. For example, the formulae of the halide ions were covered in the topic on ions, and the test for carbonate ions was previously looked at when you did the reactions of acids in chemical changes.

NAIL IT!

If you are asked to write the symbol equations for the reactions of halides and sulfate you must remember that the negative ion combines with the metal in the test reagent to form a solid.

For example, if you were writing the equation for the reaction between sodium chloride solution and silver nitrate solution the equation is as follows.

$AgNO_3(aq) + NaCl(aq) \rightarrow NaNO_3(aq) + AgCl(s)$

STRETCH IT!

At the Higher Tier you may need to write the ionic equations for these reactions.

For all three sets of ions remember that the negative ion you are testing for will react with the positive ion in the test reagent.

Ion being tested for	Testing reagent and positive ion	Ionic equation
Cl^-	Silver nitrate solution (Ag^+)	$Ag^+(aq) + Cl^-(aq) \rightarrow AgCl(s)$
Br^-	Silver nitrate solution (Ag^+)	$Ag^+(aq) + Br^-(aq) \rightarrow AgBr(s)$
I^-	Silver nitrate solution (Ag^+)	$Ag^+(aq) + I^-(aq) \rightarrow AgI(s)$
SO_4^{2-}	Barium chloride solution (Ba^{2+})	$Ba^{2+}(aq) + SO_4^{2-}(aq) \rightarrow BaSO_4(s)$
CO_3^{2-}	Hydrochloric acid (H^+)	$2H^+(aq) + CO_3^{2-}(s) \rightarrow CO_2(g) + H_2O(l)$

SNAPIT!

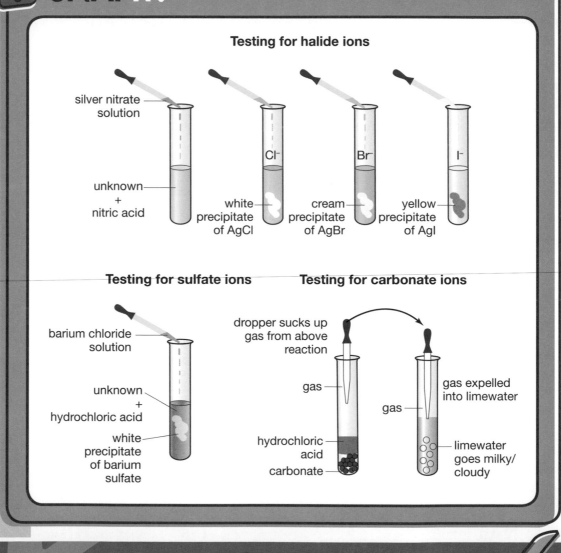

Testing for halide ions

silver nitrate solution

unknown + nitric acid

Cl^- white precipitate of AgCl

Br^- cream precipitate of AgBr

I^- yellow precipitate of AgI

Testing for sulfate ions

barium chloride solution

unknown + hydrochloric acid

white precipitate of barium sulfate

Testing for carbonate ions

dropper sucks up gas from above reaction

gas

hydrochloric acid

carbonate

gas

gas expelled into limewater

limewater goes milky/cloudy

CHECKIT!

1 What are the reagents used to test for the following ions:

    a chloride ions   **b** sulfate ions   **c** carbonate ions?

2 A solution was known to contain two negative ions. When it was tested with silver nitrate sol and nitric acid it gave a cream precipitate. The same solution also gave a white precipitate wi hydrochloric acid and barium chloride solution.

Name the two ions present.

Practical: Identifying ions in an ionic compound

This is one of the required practicals and you could be questioned on it in the exam. The practical emphasises that you carry out a practical safely and accurately, carry out chemical tests and make conclusions based on the results.

Identifying the ions in an ionic compound requires knowledge of how to test for positive and negative ions.

SNAPIT!

When testing for the halide and sulfate anions, the testing chemical is always added along with the corresponding acid. This removes any carbonate ions that would interfere with the result. For example, you always add nitric acid before adding silver nitrate solution and hydrochloric acid before adding barium chloride solution.

Stage	Test	Apparatus	Procedure
1	Flame test	Bunsen burner, tripod, gauze, heat-resistant pads, watch glass, nichrome wire, tongs, hydrochloric acid, distilled water	Solid in watch glass, add HCl to solid. Dip wire and place in roaring blue flame.
2	Sodium hydroxide	Test tubes, test-tube rack, wash bottle, droppers, sodium hydroxide solution	Make up solution in distilled water. Add NaOH solution drop by drop.
3	Negative ion tests	Test tubes, test-tube rack, wash bottle, distilled water, droppers, barium chloride solution, hydrochloric acid, silver nitrate solution, nitric acid, limewater	Add HCl to solid and if there is effervescence pass gas through limewater. If the next test is for sulfate ion then dissolve the unknown in dilute HCl and then add barium chloride solution. If it is for halide ion then dissolve in nitric acid and then add silver nitrate solution.

DOIT!

Make up a flowchart that will help you identify the cation (positive ion) and the anion (negative ion) in an ionic compound.

NAILIT!

In the exams you are likely to get a table of test results and be asked to identify the cations (positive ions) and anions (negative ions) present using these results. You will need to identify both ions to name the compound.

WORKIT!

Using the results below identify the ions present in the three compounds X, Y and Z and give the name of each compound along with its formula.

Substance	Flame test	Sodium hydroxide	Barium chloride solution	Silver nitrate solution	Hydrochloric acid
X	Crimson flame	No reaction	No reaction	No reaction	Effervescence and gas turned limewater milky
Y	No colour	White precipitate that dissolved in excess NaOH	White precipitate	No reaction	No reaction
Z	Red flame	White precipitate	No reaction	Cream precipitate	No reaction

X contains the lithium cation, Li$^+$ and the carbonate anion, CO$_3^{2-}$. The compound is lithium carbonate, Li$_2$CO$_3$.

Y contains the aluminium cation, Al^{3+} and the sulfate anion, SO$_4^{2-}$. The name of the compound is aluminium sulfate, Al$_2$(SO$_4$)$_3$.

Z contains the calcium cation, Ca^{2+} and the bromide anion, Br$^-$. The name of the compound is calcium bromide, CaBr$_2$.

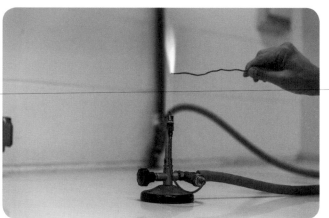

Performing a flame test

CHECKIT! ✓

1 Identify the negative ions present in A, B and C.

Substance	Barium chloride solution	Silver nitrate solution	Hydrochloric acid
A	No reaction	Yellow precipitate	No reaction
B	White precipitate	No reaction	Effervescence
C	No reaction	White precipitate	No reaction

2 How would you identify the anions and cations present in the following compounds:

a Magnesium sulfate

b Sodium carbonate

c Iron(II) chloride

d Potassium iodide?

Chemical analysis

REVIEW IT!

1 a Explain how you would carry out a flame test. Your answer should name the apparatus you would use and describe the procedure you would follow.
   b What colour flames are seen with the following compounds:
      i Sodium chloride
      ii Copper sulfate
      iii $CaCl_2$
      iv KBr?

2 What would you observe when the following actions are carried out?
   a Hydrochloric acid is added to solid calcium carbonate.
   b Nitric acid and silver nitrate solution is added to:
      i sodium iodide solution
      ii potassium chloride solution
      iii potassium bromide solution
      iv sodium chloride solution.
   c Hydrochloric acid and barium chloride solution are added to:
      i sodium sulfate solution
      ii sodium chloride solution.
   d Sodium hydroxide solution is added to:
      i magnesium sulfate solution
      ii aluminium sulfate solution
      iii iron(II) chloride solution.

3 Identify the compounds that give the results in the table below:

Substance	Flame test	Sodium hydroxide	Barium chloride solution	Silver nitrate solution
X	No colour	White precipitate	White precipitate	No reaction
Y	Yellow flame	No reaction	No reaction	Cream precipitate
Z	Lilac flame	No reaction	No reaction	White precipitate

Chemistry of the atmosphere

The composition and evolution of the Earth's atmosphere

The present composition of the Earth's atmosphere is 78% (about four-fifths) nitrogen, 21% (about one-fifth) oxygen and 1% of other gases like argon and the other noble gases, carbon dioxide, and water vapour.

This composition has evolved over billions of years.

In the early days of Earth's existence there was a lot of volcanic activity and this gave rise to an atmosphere containing lots of water vapour and carbon dioxide along with methane, nitrogen and ammonia.

As the Earth cooled, the water vapour condensed and formed the oceans, rivers and lakes.

At some stage in the Earth's history, life began and algae and photosynthetic bacteria carried out photosynthesis.

In photosynthesis glucose is made from the reaction between carbon dioxide and water. Oxygen is a by-product of this reaction.

carbon dioxide(g) + water(l) $\xrightarrow{\text{light}}$ glucose(s) + oxygen(g)

$$6CO_2(g) + 6H_2O(l) \xrightarrow{\text{light}} C_6H_{12}O_6(s) + 6O_2(g)$$

As the oxygen in the atmosphere increased because of photosynthesis, animals evolved.

As the oxygen increased, the amount of carbon dioxide was reduced by various processes. These processes were:

dissolving in water; limestone formation; photosynthesis; crude oil and natural gas formation; coal formation

Once nitrogen was formed it remained in the atmosphere because it is very unreactive. This explains its high concentration in the atmosphere.

DO IT!

Look at the apparatus in the Snap it! box. Use your notes or textbook to find out how it works.

SNAPIT!

The apparatus below is used to find the composition of air.

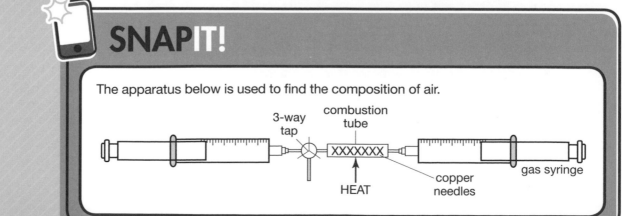

Removal process	Result
As the oceans formed carbon dioxide dissolved in the water to form insoluble solid metal carbonates.	These carbonates formed sediments so the carbon dioxide was locked in solids.
Photosynthesis uses carbon dioxide and water to make sugars such as glucose.	Carbon dioxide removed from the air.
The carbon dioxide in the oceans was converted into calcium carbonate by plankton and other marine organisms.	Their remains were compressed to form limestone.
Sometimes these dead marine organisms were covered with mud and compressed by other layers.	These remains formed crude oil and natural gas.
A similar process happened with dead land plants.	These remains formed coal as a sedimentary rock.

NAILIT!

This table shows the ways that carbon dioxide has been removed during the lifetime of the Earth.

WORKIT!

In an experiment to work out the composition of air the apparatus was the same as that shown in the Snap it! box.

$100\,cm^3$ of air was drawn into the apparatus through the 3-way tap. After heating, the volume of gas present was reduced to $79\,cm^3$.

a   What is the percentage of oxygen in air?

The volume of gas has gone down because the oxygen reacted with the copper needles.

This means that the percentage of oxygen $= \dfrac{\text{volume of oxygen}}{\text{Total volume}} \times 100\%$

$= \dfrac{21}{100} \times 100\% = 21\%$

b   What is the equation for the reaction taking place in the apparatus?

$2Cu + O_2 \longrightarrow 2CuO$

✓ CHECKIT!

1 List the main gases found in the early atmosphere of the Earth.

2 Give four ways that carbon dioxide has been removed from the atmosphere during the Earth's lifetime.

3 Give the word and symbol equations for photosynthesis.

Global warming

One of the important things about this topic is that the evidence collected by scientists has to be peer reviewed. This means that research has to be examined and tested by other scientists in the same field.

The media are also important because the interpretation they put on scientific evidence can influence public opinion and behaviour.

Most of the solar radiation reaching the Earth is shortwave UV and visible light. The energy reflected back by the Earth is longer wavelength infrared radiation.

This infrared radiation is absorbed by gases in the atmosphere which then re-emit this energy in all directions but a lot goes back to the Earth which then warms it up. This warming up is known as the greenhouse effect.

The two main greenhouse gases are carbon dioxide and methane. The rise and fall in carbon dioxide concentration in the atmosphere is followed by a rise in temperature.

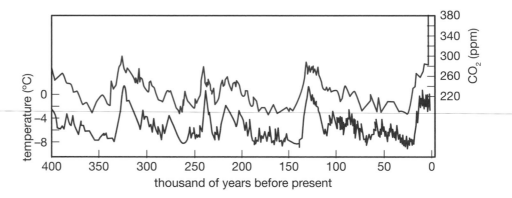

The rapid rise in carbon dioxide concentration has been blamed on the burning of fossil fuels such as crude oil and coal to generate energy.

Another cause is deforestation. Deforestation's effect is twofold because it reduces the number of trees that take up carbon dioxide by photosynthesis. Carbon dioxide is then emitted as the felled trees are burned to clear the land.

Two big sources of methane are:

• Livestock farming (from animal digestion and waste decomposition)

• Rubbish decay in landfill sites.

To stop the rise in carbon dioxide concentration humans need to drastically decrease the burning of fossil fuels.

This has proved difficult for various reasons:

• A lack of affordable alternative energy sources

• Economic growth relies on cheap energy

• Objections to the idea that global warming is caused by humans

• Lack of international co-operation.

SNAP**IT!**

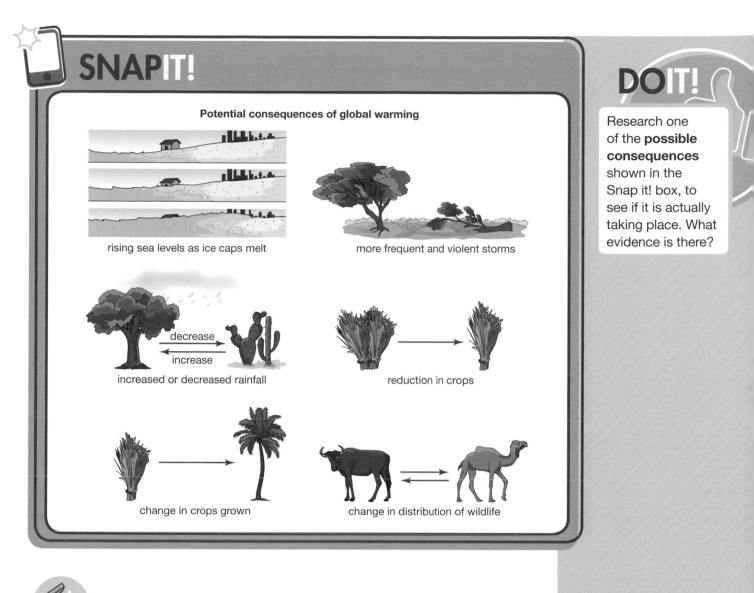

Potential consequences of global warming

rising sea levels as ice caps melt

more frequent and violent storms

decrease
increase
increased or decreased rainfall

reduction in crops

change in crops grown

change in distribution of wildlife

DO**IT!**

Research one of the **possible consequences** shown in the Snap it! box, to see if it is actually taking place. What evidence is there?

NAIL**IT!**

One question that you might have to deal with is 'Why have some countries been slow to reduce their burning of fossil fuels?' Possible answers are as follows:

- Some people think that global warming is just one part of a natural cycle.
- The same people say that evidence is still not 100% reliable.
- There is a difficulty with predicting the possible effects of global warming because different scientific models make different predictions.
- In the short term the alternatives to burning fossil fuels are more expensive.
- Lobbying by the petroleum and coal industries.
- Governments are slow to replace fossil fuels with other forms of energy.
- Inability of international bodies like the UN to get all nations to agree to reduce their use of fossil fuels.

✓ CHECK**IT!**

1 What are the two main greenhouse gases?

2 Why is the rise in carbon dioxide in the atmosphere being blamed for global warming?

3 What type of radiation is absorbed and re-emitted by greenhouse gases?

4 What is causing the rise in carbon dioxide concentration in the atmosphere?

The carbon footprint and its reduction

The carbon footprint of a product or activity is the total amount of carbon dioxide and greenhouse gases emitted over the lifetime of that product or activity.

There are a number of ways to remove or reduce the carbon footprint.

Increase the use of alternative energy supplies. For example, solar cells, wind power and wave power.

Energy conservation: reduce the amount of energy used by using energy-saving measures such as house insulation or using devices that use less energy.

In Carbon Capture and Storage (CCS) the carbon dioxide given out by power stations can be removed by reacting it with other chemicals and then stored deep under the sea in porous sedimentary rocks, especially those that used to be part of oilfields.

Carbon taxes and licences. Penalising companies and individuals who use too much energy by increasing their taxes.

By removing carbon dioxide from the air using natural biological processes, especially photosynthesis. This is done by planting trees or trying to increase marine algae by adding chemicals to the sea. This is called carbon offsetting.

Using plants as biofuels. Because plants take in carbon dioxide, when they are burned they only release the same amount of carbon dioxide, which is zero net release. This makes them carbon-neutral.

NAILIT!

You may be asked why people, companies or countries do not reduce their carbon footprints. Possible answers are as follows:

- People are reluctant to change their lifestyle. For example, still using large cars instead of smaller more fuel-efficient ones.

- Even energy-saving devices have a carbon footprint because of the processes used in the extraction of materials to make them and the energy used in their manufacture and in their disposal.

- Countries and companies still find it more economic to use lots of energy.

- Countries do not co-operate with each other.

- There is still disagreement that global warming is real and about its causes.

- People are still not sure of the facts and do not know about the possible consequences of global warming.

SNAPIT!

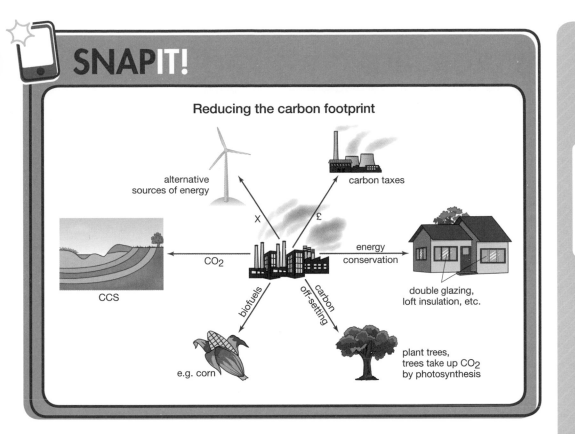

Reducing the carbon footprint

alternative sources of energy

carbon taxes

CCS

CO₂

X

£

biofuels

e.g. corn

carbon off-setting

energy conservation

double glazing, loft insulation, etc.

plant trees, trees take up CO₂ by photosynthesis

DOIT!

Look at your notes on aluminium extraction by electrolysis. Why does this process have a large carbon footprint?

CHECKIT!

1 What is meant by a carbon footprint?

2 What do you understand by the following methods of reducing our carbon footprint:

   a Carbon offsetting

   b Carbon capture and storage?

3 Give two reasons why people are not reducing their carbon footprint.

4 Photovoltaic cells and other alternative sources of energy reduce the need for burning fossil fuels but they do have a carbon footprint. Explain why.

H 5 The data below shows the savings in carbon dioxide emissions by two different methods of home insulation when a **detached house** is insulated.

Method	Reduction in carbon dioxide emissions/kg
Loft insulation	990
Cavity wall insulation	1100

   a Which of these reductions would be less for a terraced house? Explain your answer.

   b How many moles of carbon dioxide would have been emitted without insulating the loft of the detached house? ($M_r(CO_2) = 44$)

   c What is the volume of carbon dioxide that would have been emitted?

Atmospheric pollutants

An atmospheric pollutant is something that is introduced into the atmosphere and has undesired or unwanted effects.

When hydrocarbons undergo complete combustion in air they produce carbon dioxide and water along with energy.

If there is insufficient (not enough) air then incomplete combustion occurs and carbon monoxide (CO) and particulates, which are very small soot particles (mostly carbon particles (C)), form.

Carbon monoxide (CO) is a toxic gas. It combines with haemoglobin in the blood and stops it transporting oxygen around the body, especially to the brain. It can cause death.

The reaction is a reversible one:

$CO(g) + HbO_2 \rightleftharpoons O_2(g) + HbCO$

Particulates cause global dimming so that less sunlight gets through to the Earth's surface and also cause damage to the lungs.

At the high temperatures caused by combustion, nitrogen and oxygen can combine to form oxides of nitrogen. These cause respiratory problems. Eventually these oxides dissolve in water to form acid rain. Acid rain causes weathering of buildings and damages plants.

Fossil fuels contain sulfur and when this burns in air it forms sulfur dioxide (SO_2). This is a very acidic gas and when it dissolves in clouds of water vapour it causes acid rain. It also causes respiratory problems.

To reduce the formation of sulfur dioxide many petrochemicals are desulfurised.

DO IT!

Construct a mind map based on global warming, greenhouse gases and pollution.

NAIL IT!

As a rule, the oxides of all non-metals (except hydrogen) are acidic.

Make sure you are able to write the balanced equations for forming sulfur dioxide and oxides of nitrogen.

For sulfur dioxide:

$S(s) + O_2(g) \rightarrow SO_2(g)$

For oxides of nitrogen, the first oxide is nitrogen monoxide (NO):

$N_2(g) + O_2(g) \rightarrow 2NO(g)$

Another oxide of nitrogen is nitrogen dioxide (NO_2). This is very dangerous as it damages the lungs and produces acid rain:

$N_2(g) + 2O_2(g) \rightarrow 2NO_2(g)$

SNAPIT!

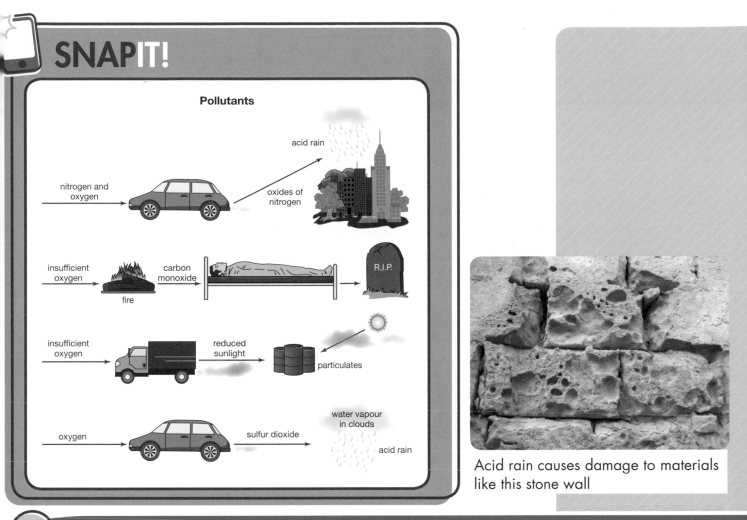

Pollutants

Acid rain causes damage to materials like this stone wall

STRETCHIT!

The equations for the formation of carbon monoxide and particulates follow the same rules as for the balancing of any other equation. You should also realise that you are allowed to use ½ molecules of oxygen in these equations and water is always formed as the other product.

The burning of methane is the simplest example of the combustion of a hydrocarbon. Just for comparison, the equation for the complete combustion of methane is as follows:

$$CH_4(g) + 2O_2(g) \rightarrow CO_2(g) + 2H_2O(l)$$

The equation for carbon monoxide formation is:

$$CH_4(g) + 1½O_2(g) \rightarrow CO(g) + 2H_2O(l)$$

For particulates (which are carbon):

$$CH_4(g) + O_2(g) \rightarrow C(g) + 2H_2O(l)$$

Notice for both of these less oxygen is used up and the methane undergoes incomplete combustion.

CHECKIT!

1 Name four pollutants formed by burning fossil fuels and for each one give its effects.

2 Explain why a coal or wood fire should always be in a well ventilated room.

H 3 a i Write the equation for the complete combustion of methane to give carbon dioxide.

   ii Write the equation for the incomplete combustion of methane to give carbon monoxide.

1   State the main gases in the atmosphere of the early Earth.

2   List five ways by which the amount of carbon dioxide in the atmosphere has been reduced.

3   Describe the composition of the present atmosphere of the Earth.

4   Name two greenhouse gases and explain how they cause the greenhouse effect.

5   Describe the evidence that rising levels of carbon dioxide are causing global warming.

6   Why might some areas expect flooding as a result of global warming?

7   Explain what is meant by the term 'carbon footprint'.

8   Describe five ways by which the carbon footprint of an organisation or individual person can be reduced.

9   Explain why some countries might be reluctant or slow to reduce their carbon footprint.

10   What is a pollutant?

11   Carbon monoxide is a pollutant.

    a i   What is the formula of carbon monoxide?

        ii How is it formed?

        iii What are its undesired effects?

    b Oxides of nitrogen are pollutants.

        i   How are they formed?

        ii What are their undesired effects?

12   Explain why sulfur should be removed from petrol.

Using resources

Finite and renewable resources, sustainable development

The natural resources used by chemists to make new materials can be divided into two categories – finite and renewable. Finite resources will eventually run out. Examples are fossil fuels and various metals even though we are able to get some valuable materials from places like the oceans.

Renewable resources are ones that can be replaced at the same rate as they are used up. They are derived from plant materials. An example is ethanol, which is made from sugar from fermentation. Ethanol can be used as fuel for cars instead of petrol which is extracted from the finite resource, crude oil.

Many of the Earth's natural resources are running out and if they are used at the current high rates they will be depleted (used up) very soon. In order to increase the lifetime of these finite resources the industry has to develop processes that increase the lifetime of natural resources.

Sustainable development meets the needs of present development without depleting natural resources. In a sustainable process:

- there is a high yield
- there are few waste products
- there is very little impact on the environment and the products should not harm the environment.

DO IT!

The list below shows different methods of waste management.

Put them in order from 'Worst' to 'Best' method in terms of increasing the sustainability of a product.

A Burning product to produce energy

B Reuse the product

C Placing in landfill

D Design product differently to prevent waste

E Recycle or compost the product

SNAP IT!

Sustainable processes should do the following:

- have reactions with high atom economy
- use renewable resources
- have as few steps as possible
- use catalysts.

MATHS SKILLS

One of the main concerns related to sustainability is the number of years left before certain finite resources are exhausted.

The remaining reserves are usually in very large amounts and they are usually expressed in millions (10^6), billions (10^9) and trillions (10^{12}).

In order to answer these questions you should be able to manipulate numbers by expressing them in standard form.

WORKIT!

In 2013 the known USA reserves of natural gas were 646 trillion m^3.

At the present rate the consumption is 765 billion m^3 per annum.

Calculate how many years the USA has left before its natural gas is depleted.

The reserves $= 646 \times 10^{12} m^3 = 6.46 \times 10^{14} m^3$

The consumption per annum $= 765 \times 10^9 m^3 = 7.65 \times 10^{11} m^3$

Therefore the number of years left $=$ reserves/consumption per annum

$= 6.46 \times 10^{14} / 7.65 \times 10^{11}$

$= 844$ years

An oil pump in a field of rapeseed

CHECKIT! ✓

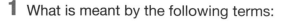

1 What is meant by the following terms:

   a Finite resource

   b Renewable resource

   c Sustainable development?

2 Describe four characteristics of a sustainable process and for each one explain why it increases the sustainability of the process.

3 The USA has 17.7 billion metric tonnes of coal reserves remaining. The annual consumption of coal in the USA is 175 million metric tonnes per annum. How long will it be before the USA runs out of its own coal?

What could be done to extend this time?

Life cycle assessments (LCAs)

A life cycle assessment is an analysis of the environmental impact of a product at each stage of its lifetime, from its production all the way to its disposal.

The stages in a life cycle assessment are as follows:

The extraction/production of raw materials.

The production process – making the product, the packaging and any labelling.

How the product is used and how many times it is used.

The end of the life of the product – how is it disposed of at the end of its lifetime? Is it recycled?

The diagram in the Snap it! box shows a typical life cycle assessment.

The use of energy, resources and waste production can be calculated or measured reasonably accurately.

On the other hand, pollution effects are more difficult to measure or calculate.

LCAs can be used by companies to avoid unnecessary generation of waste and lead them to rethink their procedures.

DO IT!

Draw a table that summarises the advantages and disadvantages to the environment of using paper and plastic bags.

SNAP IT!

What happens to product after its lifetime comes to an end
- Products that can be recycled easily
- Products that produce low toxic products when disposed of

Raw materials
- Either extract raw materials
  or
- Use recycled materials

Manufacturing
- Measures to prevent air and water pollution
- Energy conservation
- Use of recycled raw materials and components

Usage
- Products that use less power
- Products that use less water and chemicals

Distribution of product
- Simple cheap packaging
- Short delivery distances

WORKIT!

What is a more sustainable product – a paper bag or a plastic bag?

Paper bags come from trees – a renewable resource. Plastic (polyethylene) bags are made from ethene which is produced during cracking of petrochemicals – a finite resource.

During the production stage paper has a greater impact on the environment than plastic bags. Paper bag production consumes much <u>more water</u>, produces <u>more acidic greenhouse gases</u> and has a greater global warming effect.

Paper bags are damaged by water and <u>less easily reused</u>.

Paper bags are heavier and after use generate 5 times <u>more solid waste</u> than plastic bags.

These results show that plastic bag production has less unwanted effects than for paper bags.

151

CHECKIT!

1 What is a life cycle assessment?

2 State the four stages/parts of a life cycle assessment.

3 The table below shows the greenhouse emissions in grams of greenhouse gas per kilowatt hour of power produced for **four** ways of producing electricity. The measurements are made over the lifetime of the generating device.

Method of electricity generation	Greenhouse gas emissions (in g/kWh) over the lifetime of the device
Silicon photovoltaic cell	45
Coal power station	900
Natural gas power station	420
Nuclear power station	30

a Comment on, and explain the values for, greenhouse emissions of the coal and natural gas power stations.

b Comment on, and explain the values for, the photovoltaic cell and the nuclear power station.

4 The flowcharts below show the greenhouse emissions as a percentage of the total for stages in the lifetimes of a photovoltaic cell and a coal power station.

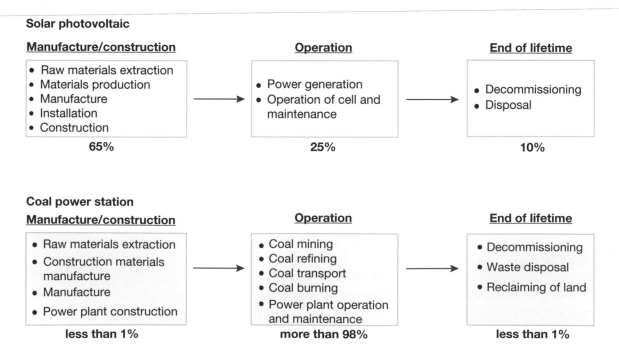

Solar photovoltaic

Manufacture/construction	**Operation**	**End of lifetime**
• Raw materials extraction • Materials production • Manufacture • Installation • Construction	• Power generation • Operation of cell and maintenance	• Decommissioning • Disposal
65%	**25%**	**10%**

Coal power station

Manufacture/construction	**Operation**	**End of lifetime**
• Raw materials extraction • Construction materials manufacture • Manufacture • Power plant construction	• Coal mining • Coal refining • Coal transport • Coal burning • Power plant operation and maintenance	• Decommissioning • Waste disposal • Reclaiming of land
less than 1%	**more than 98%**	**less than 1%**

a Explain why the bulk of the emissions for the photovoltaic cell come at the beginning.

b For the coal power station only 1% of the greenhouse emissions come at the beginning of its lifetime and only 1% comes at the end of its lifetime. Explain these relatively small numbers.

Alternative methods of copper extraction

Copper is **extracted** from copper-rich **ores** by **smelting** and **electrolysis**.

In smelting, use is made of the fact that copper is less reactive than carbon so the ore is roasted with **carbon**. For example, malachite, an ore which is mostly copper(II) carbonate, is first decomposed to copper(II) oxide.

$$CuCO_3(s) \rightarrow CuO(s) + CO_2(g)$$

The copper(II) oxide is then **reduced** to copper by the carbon.

$$2CuO(s) + C(s) \rightarrow Cu(s) + CO_2(g)$$

If the ore contains mainly copper(II) sulfide (CuS) then reaction with oxygen produces impure copper and sulfur dioxide.

$$CuS(s) + O_2(g) \rightarrow Cu(s) + SO_2(g)$$

Some of the copper(II) sulfide produces copper(II) oxide which can then be reduced using carbon as shown above.

The other method is electrolysis. In this method sulfuric acid is added to give copper(II) sulfate solution. Electrolysis gives pure copper at the **cathode**.

If the copper produced by smelting is impure then this impure copper is made the **anode** in electrolysis. It dissolves to give copper(II) ions which are then deposited on the cathode as pure copper.

copper anode $\xrightarrow{\text{loses 2 electrons}}$ copper(II) ions in solution $\xrightarrow{\text{gains 2 electrons}}$ copper on cathode

- If the copper ores are low-grade ores (low in copper) then it is **uneconomical** to extract copper from them using the usual methods so alternatives are used. **Bioleaching** is a process where **bacteria** use copper sulfide as a source of energy and separate out the copper, which can be filtered off from the liquid produced. Bioleaching is slow but uses only about 30 to 40% of the energy used by copper smelting.

- **Phytomining** takes advantage of plants that take up copper from slag heaps that contain low-grade copper ores. The copper concentrates in the plant and is left in the ash when the plants are burned. Sulfuric acid is added to the ash to give copper(II) sulfate solution.

Phytomining is an environmentally friendly but slow method of extraction.

Copper can be extracted from this copper(II) sulfate solution by adding scrap iron. The iron displaces pure copper from the solution.

$$Fe(s) + CuSO_4(aq) \rightarrow Cu(s) + FeSO_4(aq)$$

DO IT!

Record an MP3 describing the extraction of copper by smelting, bioleaching and phytomining. Alternatively explain the processes to a revision partner.

NAIL IT!

H

The electrolysis of copper sulfate can use different electrodes. If inert electrodes (electrodes that do not take part) are used then copper ions from the solution are discharged at the cathode as copper.

If a copper anode is used then the copper at the anode dissolves to give copper(II) ions (Cu^{2+}). These copper ions gain electrons as they are discharged at the cathode to give pure copper. This means that copper can be purified by making impure copper the anode in electrolysis.

SNAPIT!

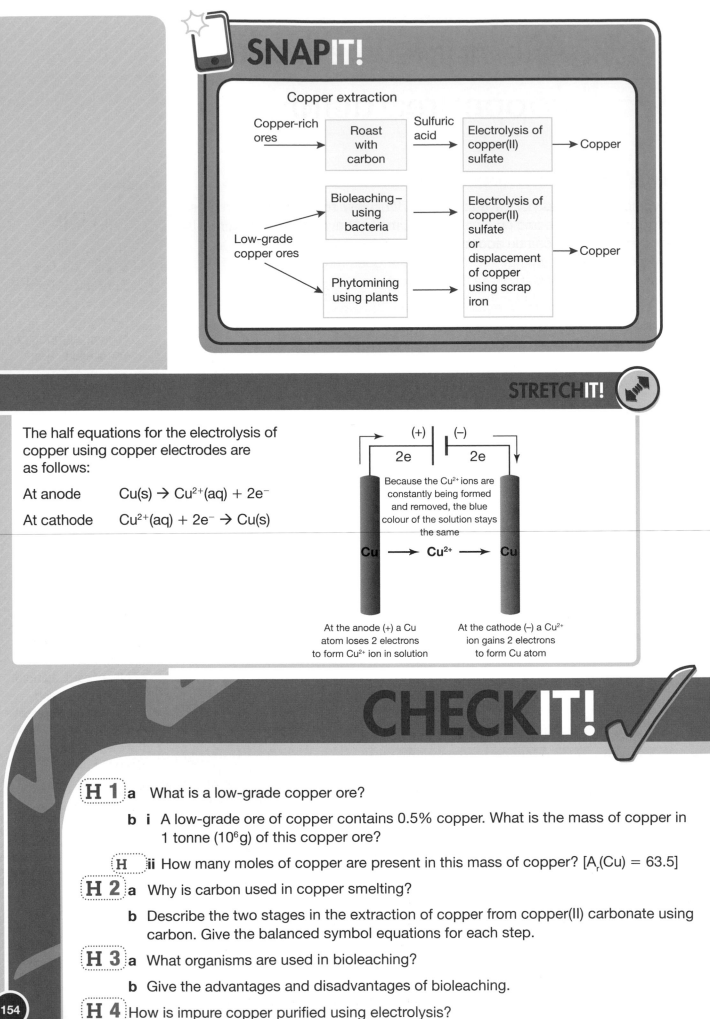

Copper extraction

Copper-rich ores → Roast with carbon → Sulfuric acid → Electrolysis of copper(II) sulfate → Copper

Low-grade copper ores → Bioleaching – using bacteria → Electrolysis of copper(II) sulfate or displacement of copper using scrap iron → Copper

Low-grade copper ores → Phytomining using plants → Electrolysis of copper(II) sulfate or displacement of copper using scrap iron → Copper

STRETCHIT!

The half equations for the electrolysis of copper using copper electrodes are as follows:

At anode $Cu(s) \rightarrow Cu^{2+}(aq) + 2e^-$

At cathode $Cu^{2+}(aq) + 2e^- \rightarrow Cu(s)$

(+)   (–)
2e   2e

Because the Cu^{2+} ions are constantly being formed and removed, the blue colour of the solution stays the same

Cu → Cu^{2+} → Cu

At the anode (+) a Cu atom loses 2 electrons to form Cu^{2+} ion in solution

At the cathode (–) a Cu^{2+} ion gains 2 electrons to form Cu atom

CHECKIT!

H 1 **a** What is a low-grade copper ore?

   b **i** A low-grade ore of copper contains 0.5% copper. What is the mass of copper in 1 tonne (10^6 g) of this copper ore?

   H **ii** How many moles of copper are present in this mass of copper? [$A_r(Cu) = 63.5$]

H 2 **a** Why is carbon used in copper smelting?

   b Describe the two stages in the extraction of copper from copper(II) carbonate using carbon. Give the balanced symbol equations for each step.

H 3 **a** What organisms are used in bioleaching?

   b Give the advantages and disadvantages of bioleaching.

H 4 How is impure copper purified using electrolysis?

Making potable water and waste water treatment

Potable water is water that is safe to drink.

This type of water does not have to be pure. It usually contains small concentrations of salts and no microbes. It can also have chemicals added. For example, fluoride can be added to reduce dental decay in children. In areas where water is plentiful, the impure water goes through a series of processes to make it potable (see Snap it! box).

In arid countries sea water is made potable by distillation or reverse osmosis. Israel, for example, gets very little rain and uses reverse osmosis to get about 70% of its water. These methods use a lot of energy.

SNAPIT!

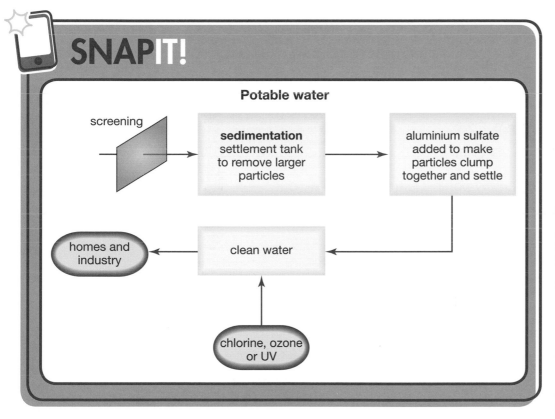

Potable water

Large amounts of waste water are produced by industry and domestic consumers.

Before it is returned to the environment this water is treated using a number of processes including filtration/screening, sedimentation or settling.

Large pieces of grit and soil are separated by screening or filtration and the remaining liquid is passed into settling tanks.

After sedimentation the sludge obtained is treated using anaerobic digestion (carried out in the absence of oxygen). This produces methane which can be used as a fuel to run the sewage plant and a solid that can be used as a fertiliser or as a fuel. The liquid or effluent is then treated using aerobic digestion (in the presence of oxygen) and then returned to the environment.

NAILIT!

This topic is closely aligned with that of sub-topics Mixtures and compounds and Identifying ions in an ionic compound. You should revise these together as the purity of water is an important concept. You should know how to make sure that your water is pure or impure.

DOIT!

Write the different steps for water purification on same-sized cards or bits of card. Jumble or shuffle them and put them in the correct order. Compare with the diagrams in the Snap it! boxes.

Repeat for the treatment of waste.

155

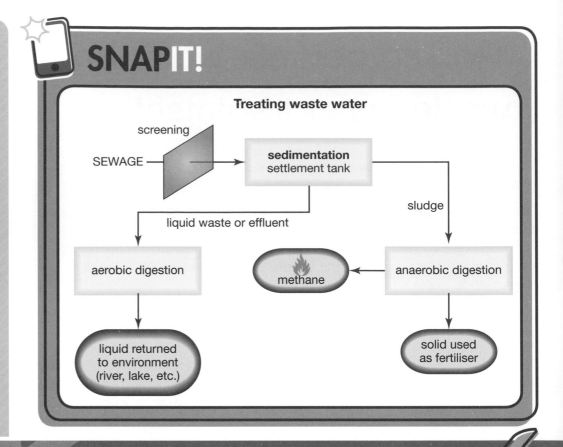

SNAPIT!

Treating waste water

CHECKIT! ✓

1 a What is meant by potable water?

b How could you show that potable water contained:

   i dissolved solids   **ii** chloride ions?

c i Describe a chemical test for pure water.

   ii Describe a test for pure water based on a physical property.

2 Which chemical technique does screening resemble?

3 What do you understand by the following terms:

   a anaerobic   **b** aerobic   **c** sedimentation?

4 a Why would the production of methane in the anaerobic digestion of solid waste increase the sustainability of water treatment processes?

b Give any other ways that the process is made more sustainable.

5 Some water can be extracted from underground rocks such as limestone. What stage in the purification process may be unnecessary if this is the source of water?

H 6 A sample of water was thought to contain **chloride** <u>and</u> **bromide** ions. The water would not pass for human consumption if bromide ions were present. Two students were given the task of showing that bromide ions were present.

Student I suggested adding silver nitrate solution along with nitric acid.

Student II suggested passing chlorine gas through the water.

a Explain why the process suggested by student I would not give a clear-cut result.

b Explain why the method suggested by student II would work. What observation would confirm the presence of bromide ions?

Ways of reducing the use of resources

If we do not reduce our use of different materials then the sources of these materials will run out.

Extraction of metals from their ores requires lots of energy and therefore will also further deplete the reserves of fossil fuels still available.

Extraction also leads to the creation of more waste and mining the metal ores has bad impacts on the environment.

- Glass bottles can be reused or the glass can be crushed and melted to be reformed. This saves energy and conserves resources that are used to make the glass.

- Aluminium extraction requires lots of electrical and heat energy and the ore of aluminium (bauxite) is running out. Bauxite mining and concentration has a bad impact on the environment. Recycling aluminium saves 95% of the energy used in extraction and produces 95% less greenhouse gases.

- Separating iron from other metals is relatively easy because it is magnetic. Recycling iron and steel saves a lot of energy (and because of this a lot of fossil fuels) and reduces the emission of greenhouse gases and pollutants. Scrap iron is also used to help in the production of steel.

- Plastics can be sorted and then recycled or incinerated. They can also be cracked to give hydrocarbon fuels and alkenes.

DO IT!

Choose a material and sketch out a poster to persuade people to recycle the material.

SNAP IT!

Reusing and recycling reduces

- use of fossil fuels
- greenhouse gas emissions
- mining of ores
- negative impact on environment

NAIL IT!

Revise the extraction of aluminium at the same time as you revise this topic – the savings made by recycling aluminium will make more sense.

CHECK IT!

1 Give at least three advantages of recycling metals and glass.

2 a Why is so much aluminium recycled?

   b How could aluminium be separated from iron in a recycling plant?

3 Sometimes waste plastics can be cracked to give alkenes as one of the products. Why can this be thought of as a way of recycling?

Rusting

Corrosion occurs when a metal reacts with substances in the environment. This leads to weakening of the metal.

Aluminium does not corrode because it has a protective layer of aluminium oxide that does not flake off.

The corrosion of iron is called rusting.

Rust is a form of hydrated iron(III) oxide ($Fe_2O_3.nH_2O$). This is why both water and air are required for rusting to take place.

Iron(s) + oxygen(g) + water(l) → hydrated iron(III) oxide(s)

When iron(III) oxide (Fe_2O_3) forms it flakes off, exposing iron underneath to more rusting.

This generator is covered in rust

Because iron is so widely used, rusting is very costly.

There are two ways to stop rusting taking place.

Barrier methods. These stop air and water getting to the iron. Examples are painting, greasing/oiling and coating with tin or plastic (for food cans).

Using sacrificial metals. These are metals more reactive than iron and they react instead of the iron. Coating with zinc is called galvanising. If the zinc is scratched the iron will still not rust because the zinc reacts instead of the iron.

Sacrificial metals are used to stop rusting on ships. Cleaning and repainting is costly so blocks of the sacrificial metal, magnesium, are placed on the side of the ship and when they are finished they are simply replaced.

SNAPIT!

Experiments on rusting

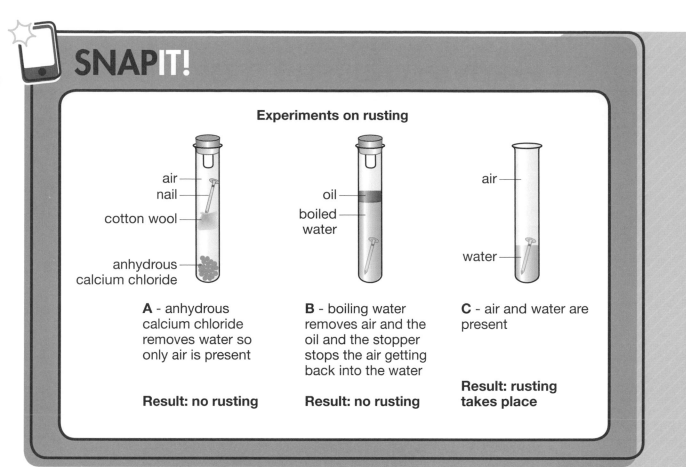

air
nail
cotton wool
anhydrous
calcium chloride

oil
boiled
water

air

water

A - anhydrous calcium chloride removes water so only air is present

Result: no rusting

B - boiling water removes air and the oil and the stopper stops the air getting back into the water

Result: no rusting

C - air and water are present

Result: rusting takes place

Rust can cause damage to ships

DOIT!

Sketch out a plan for an experiment that tests what effect less reactive metals have on the speed of rusting.

CHECKIT!

1 What is corrosion?

2 Apart from iron, which substances are needed for rusting to occur?

3 The formula of rust can be written as $Fe_2O_3.2H_2O$. Complete the equation below to show its formation.

$$\_Fe(s) + \_O_2(g) + \_H_2O\ (l) \rightarrow \_Fe_2O_3.2H_2O(s)$$

4 How does painting stop rusting taking place?

5 How does magnesium stop rusting taking place?

Alloys as useful materials

Pure metals are usually soft because the layers of metal ions can easily slide over each other without disrupting the structure. Mixing a metal with other metals stops the layers sliding over each other and the metal is made harder.

Alloys are mixtures of metals. Steels are mixtures of iron with different amounts of carbon and other metals.

In steel the amount of carbon with the iron affects the properties of the steel.

Steels are harder if the amount of carbon is increased.

- Mild steel which is used for car bodies and pipes has a low percentage of carbon because the steel needs to shaped and bent.

- High-carbon steel is very hard and this makes it ideal for use in hammers and drills. Even though this is high carbon it is only about 1–1.5%.

The other metals added to steel depend on the properties required.

If you want steel that is hard and does not corrode then chromium and nickel are added. This type of steel is stainless steel and used to make cooking utensils, knives and forks, etc.

Gold is a very soft metal and to make it more hardwearing for objects like rings it is alloyed with metals like copper, silver and zinc.

Pure gold is 24 carat. A 12 carat gold object is 50% gold and a 9 carat gold object is about 37.5% gold.

Aluminium alloys are used for aeroplane bodies because they have a high strength to weight ratio. The metals added to the aluminium are magnesium and copper.

Copper is very soft. Brass is an alloy of copper and zinc which is harder than copper but still malleable. Brass is used for bathroom fittings and musical instruments.

Bronze is an alloy of copper and tin used to make statues.

DO IT!

Sketch diagrams of the metal particles in a pure metal and an alloy. Use these diagrams to explain why steels get harder as the percentage of carbon increases.

WORKIT!

A sample of brass contained 25.4 g of copper and 6.5 g of zinc.

What is the percentage composition of the brass in terms of moles? A_r of copper = 63.5 and A_r of zinc = 65.

Note we use moles to work out the percentages because this gives a measure of the ratio of metal atoms in the alloy.

Number of moles of copper = $mass/A_r$ = 25.4/63.5 = 0.400 mol;
Number of moles of zinc = 6.5/65 = 0.100 mol.

Percentage of copper = $\dfrac{\text{no. of moles of copper}}{\text{total no. of moles}} \times 100\% = \dfrac{0.400}{0.500} \times 100 = 80\%$;

Percentage of zinc = 20%.

SNAPIT!

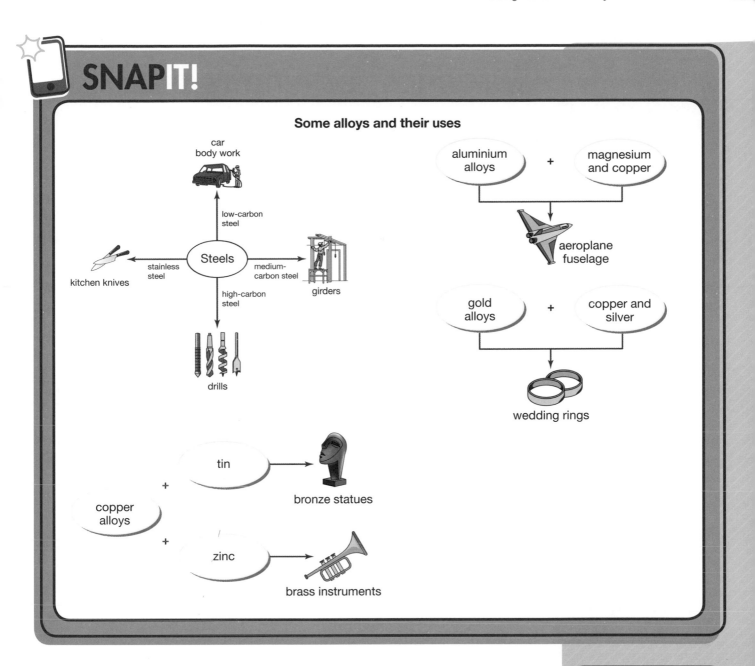

Some alloys and their uses

MATHS SKILLS

Working out molar amounts and percentages in **H** alloys. Remember $n = m/A_r$.

CHECKIT!

1 What is an alloy?

2 What is the effect of decreasing the amount of carbon in steel?

3 What metals make up the following alloys:

  a brass

  b bronze

  c stainless steel?

H 4 A sample of bronze was found to contain 14.3 g of tin and 55.9 g of copper. What is the percentage composition of the bronze in terms of moles? [A_r(Cu) =63.5 and A_r(Sn) = 119]

Ceramics, polymers and composites

DO IT!

Look up other examples of composites. If you have a squash or tennis racquet then what is the matrix and what is the reinforcement?

NAIL IT!

Make sure you understand that intermolecular forces only work over short distances and are weak but increase when there are more points of contact between the molecules and polymer chains.

In LDPE the branching stops the chains getting close together so very few intermolecular forces are able to form between the chains. This means that the polymer has a low density and the low attractions between the chains means that they can slide over each other and so the polymer is very flexible and stretches easily.

Glass is formed when sand (silicon dioxide) is melted and then rapidly cooled.

Adding other compounds to the sand when forming the glass gives different types of the material.

Soda-lime glass is used for making windows and is made from sand, sodium carbonate and limestone.

Borosilicate glass is made from sand and boron trioxide. It has a higher melting point than soda-lime glass and it is used for glass objects that can be heated, e.g. test tubes, flasks, etc.

The structure of glass is disordered and resembles the structure of a liquid. Ceramics consist of metal ions and covalent structures arranged in layers.

If substances like clay are mixed with water these layers slide over each other. After heating the water is removed and strong bonds are formed between the layers.

Ceramics are hard, brittle and electrical insulators.

The properties of polymers depend on the monomers from which they are made and the production conditions used.

There are several types of polyethylene (PE). Low-Density PolyEthylene (LDPE) is formed at high pressure and high temperature. The biggest use of LDPE is plastic bags.

High-Density PolyEthylene (HDPE) is made using special catalysts and uses lower temperatures and pressures. Examples of uses are corrosion-resistant pipes, milk jugs and plastic wood substitutes.

Thermosoftening polymers are made of individual polymer chains. When heated these chains are easily separated and the polymer melts.

Thermosetting polymers have chains that are linked by covalent bonds. When these polymers are heated they do not melt.

Composites are materials made from two or more materials that have different properties and when combined produce a material with different properties from the constituent materials.

Many composites consist of a material that provides strength or reinforcement and another that provides the supporting matrix.

Examples of composites are:

Reinforced concrete – the matrix is the cement and the reinforcement is gravel and steel wires.

Fibre glass – strands of glass fibre provide the reinforcement and a plastic resin provides the matrix.

SNAPIT!

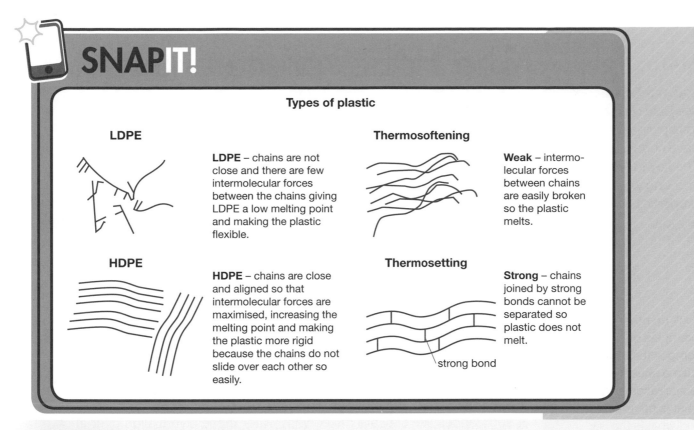

Types of plastic

LDPE

LDPE – chains are not close and there are few intermolecular forces between the chains giving LDPE a low melting point and making the plastic flexible.

HDPE

HDPE – chains are close and aligned so that intermolecular forces are maximised, increasing the melting point and making the plastic more rigid because the chains do not slide over each other so easily.

Thermosoftening

Weak – intermolecular forces between chains are easily broken so the plastic melts.

Thermosetting

Strong – chains joined by strong bonds cannot be separated so plastic does not melt.

strong bond

WORKIT!

The table below shows the physical properties of an alloy, stainless steel, and a ceramic, silicon carbide.

Property	What does the property tell you?	Silicon carbide (SiC)	Stainless steel
Hardness/GPa	Resistance to wear and can withstand impact better	22	5
Toughness/MPa per m^2	Resistance to fracturing and chipping	4	220
Specific gravity	Density compared with water	3.2	7

Why is stainless steel better for knives and scissors than silicon carbide?

Stainless steel has a much higher toughness so it will not chip easily when used.

CHECKIT!

1 What are the substances used to make borosilicate glass?

2 Why is the structure of glass similar to that of a liquid?

3 a What do the letters LDPE and HDPE stand for?

  b Describe the packing of the chains in HDPE.

  c Why does this make HDPE a more rigid material than LDPE?

4 a Explain what is meant by a composite material.

  b Explain how reinforced concrete works as a composite material.

5 Explain why thermosetting plastics do not melt.

6 Using the table of properties shown above in the worked example explain why:

  a silicon carbide is better for grinding tools

  b silicon carbide is used rather than stainless steel in body armour.

The Haber process

Ammonia has the formula NH_3. It is an important starting material for the production of nitric acid and fertilisers.

The Haber process is used to manufacture ammonia from the elements nitrogen and hydrogen.

The equation for the reaction is:

$$N_2(g) + 3H_2(g) \rightleftharpoons 2NH_3(g)$$

The reaction is reversible and the forward reaction is exothermic. This means that the reverse reaction is endothermic.

In the industrial process the nitrogen comes from the fractional distillation of liquid air and the hydrogen comes from natural gas.

The conditions used are a temperature of 450°C, a pressure of 200–250 atmospheres and an iron catalyst.

In the industrial process the conversion to ammonia is about 20–25%. The ammonia that is formed is cooled, condenses and is then run off as a liquid.

Unreacted nitrogen and hydrogen is recycled.

Because it is exothermic, the forward reaction would be favoured by a lowering the temperature and we get more ammonia if the temperature is low.

H

The problem is that a low temperature would make the reaction slow. To overcome this problem a compromise is arrived at and a temperature of 450°C is used. This gives a reasonable rate of reaction and a passable yield.

The formation of ammonia means that the number of gas molecules goes down from four to two. If the pressure is raised the chemical system will try and lower it by making fewer gas molecules and this means that formation of ammonia is favoured by a high pressure.

The pressure used is about 200 to 250 atmospheres. Higher pressures would give a better yield but are expensive.

An iron catalyst is used to speed up the reaction. A catalyst will not affect the position of equilibrium.

In the industrial process the amounts of nitrogen and hydrogen are in the ratio 1 nitrogen to 3 hydrogens as in the equation.

H

DO IT!

Write down the different ways that money is saved in the Haber process.

SNAP IT!

The Haber Process

Nitrogen from air

Hydrogen from natural gas

→ Heater and compressor

Unreacted nitrogen and hydrogen is recycled

Catalyst chamber (iron catalyst)

$N_2(g) + 3H_2(g) \rightleftharpoons 2NH_3(g)$

250 atmospheres pressure at 450°C

Condenser

Ammonia run off as liquid

MATHS SKILLS

You need to be able to read data from graphs.

WORKIT!

What is the percentage conversion to ammonia at 500°C and 300 atmospheres pressure?

There are 4 lines so use the correct one.

The answer is 19%.

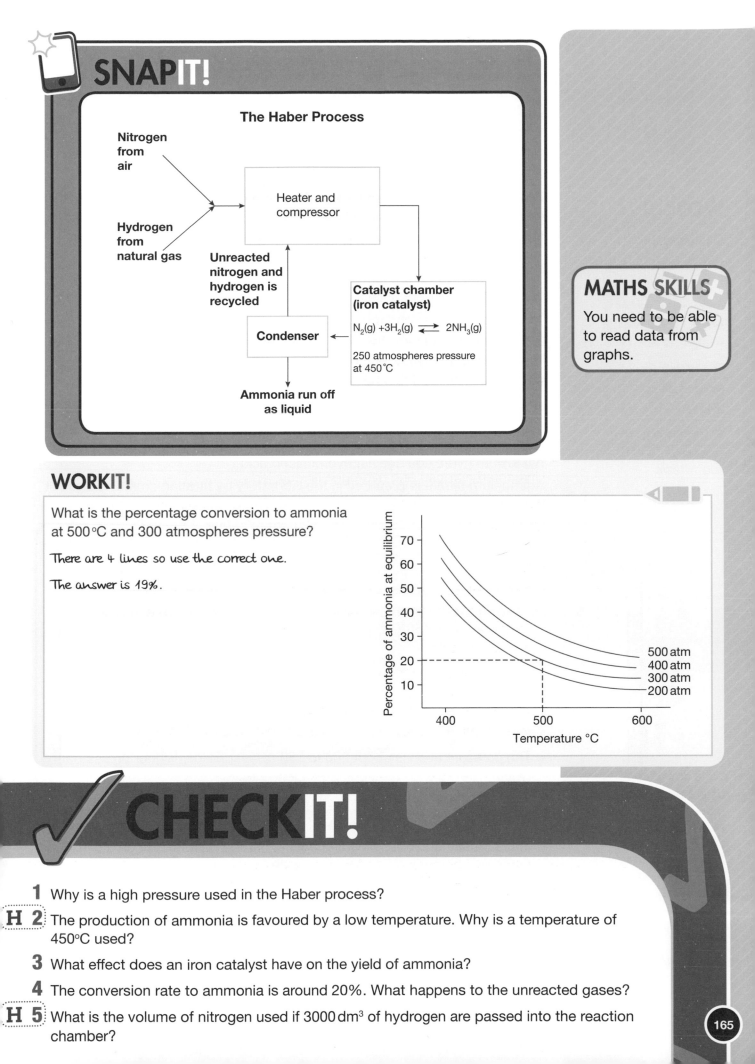

Percentage of ammonia at equilibrium vs Temperature °C

500 atm
400 atm
300 atm
200 atm

CHECK IT!

1 Why is a high pressure used in the Haber process?

H 2 The production of ammonia is favoured by a low temperature. Why is a temperature of 450°C used?

3 What effect does an iron catalyst have on the yield of ammonia?

4 The conversion rate to ammonia is around 20%. What happens to the unreacted gases?

H 5 What is the volume of nitrogen used if 3000 dm³ of hydrogen are passed into the reaction chamber?

Production and uses of NPK fertilisers

WORKIT!

Calculate the percentage of nitrogen in ammonium nitrate.

Ammonium nitrate is NH_4NO_3. The mass of nitrogen is $2 \times 14 = 28$

The relative formula mass $= 14 + 4 + 14 + 48 = 80$

The percentage of nitrogen $= 28/80 \times 100 = 35\%$

Plants need compounds of nitrogen (N), phosphorus (P) and potassium (K) for growth and carrying out photosynthesis.

Fertilisers containing these three elements are called NPK fertilisers.

Ammonia from the Haber process is oxidised to form nitric acid (HNO_3) which is then reacted with ammonia (NH_3) to give ammonium nitrate (NH_4NO_3). This is a good fertiliser because it is water-soluble and contains lots of nitrogen.

To get phosphorus, phosphate rock is treated with nitric acid to form phosphoric acid (H_3PO_4) and calcium nitrate ($Ca(NO_3)_2$.

The phosphoric acid is reacted with ammonia to give ammonium hydrogen phosphate ($(NH_4)_2HPO_4$) which is soluble.

To give NPK fertilisers the ammonium nitrate and ammonium hydrogen phosphate are mixed with potassium chloride. Potassium chloride is obtained by mining.

In the laboratory potassium chloride would be prepared by titration of potassium hydroxide against hydrochloric acid.

Ammonium nitrate is prepared in the laboratory by titrating ammonia against nitric acid.

SNAPIT!

The flowchart shows the materials used in the **industrial preparation of an NPK fertiliser.**

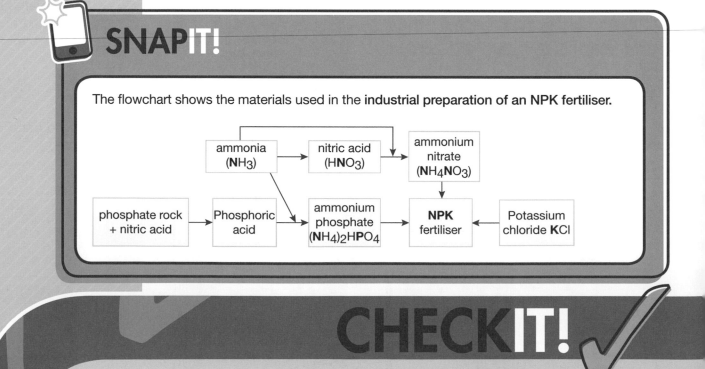

CHECKIT!

1 What is an NPK fertiliser?

2 Why is ammonium nitrate a good fertiliser?

3 Why do you think fertiliser production is on the same site as the Haber process?

4 Give the equation for the formation ammonium hydrogen phosphate fro ammonia and phosphoric acid.

5 Calculate the percentage of phosphor in ammonium hydrogen phosphate.

Practical: Analysis and purification of a water sample

This is one of the required practicals and you could be questioned on it in the exam. The practical emphasises that you carry out a practical safely and accurately and safely use a range of equipment to purify and/or separate chemical mixtures including evaporation and distillation.

Practical Skills

Questions you should be able to answer:

- What apparatus can I use?
- How can you test the water sample for purity?
- What chemical test should be used?

NAILIT!

Questions you could ask yourself:

- How can you show that there is a dissolve solid in the solution you are given? What apparatus would you use?
- What tests would you use to identify any dissolved solids?
- How do the pieces of apparatus shown in the diagrams work to purify the water?
- How would you show that the purified water was indeed pure?

SNAPIT!

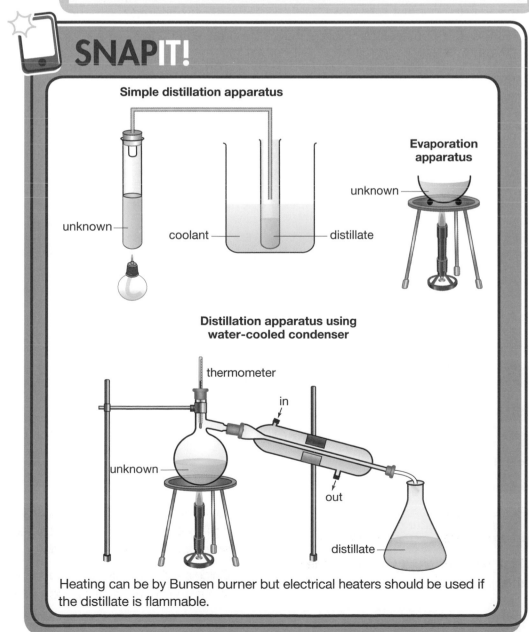

Simple distillation apparatus

unknown — coolant — distillate

Evaporation apparatus

unknown

Distillation apparatus using water-cooled condenser

thermometer

in

unknown

out

distillate

Heating can be by Bunsen burner but electrical heaters should be used if the distillate is flammable.

167

WORKIT!

A sample of water was evaporated to dryness and a residue was obtained which was white in colour.

When the residue was tested using the flame test no colour was obtained.

Samples of the water were added to separate test tubes and the following results were obtained:

1 On addition of sodium hydroxide the solution gave a white precipitate.

2 A separate sample gave no precipitate with silver nitrate solution but did give a dense white precipitate with barium chloride solution.

3 After using a simple distillation apparatus to purify the water, the distillate gave very faint white precipitates with both sodium hydroxide and separately with barium chloride solution.

4 When a condenser and flask were used to purify the water, the distillate gave no precipitates with the sodium hydroxide solution or barium chloride solution.

What do these results show?

The white residue showed that the water was impure and did contain a dissolved solid.

The white colour of the residue indicates that the original solution did not contain a transition metal, otherwise it would have been coloured.

The result with sodium hydroxide solution shows that the original solution contained magnesium ions.

The result with silver nitrate solution showed that no halide ions were present but the result with barium chloride solution did show the presence of sulfate ions.

These results show that the dissolved substance in the water was magnesium sulfate.

After using the simple distillation apparatus the fainter precipitates show that less of the dissolved solid was present but that the purification process was not completely successful.

After distillation using the condenser no precipitate with either sodium hydroxide or barium chloride solution showed there wasn't any dissolved substance in the water. This showed that purification was successful.

CHECKIT! ✓

1 A sample of water turned universal indicator paper red. It gave no residue when it was evaporated to dryness. When tested with silver nitrate solution it gave a white precipitate.

    a i What apparatus could you use instead of the UI paper?

      ii What is the advantage of using this apparatus over the UI paper?

    b What is the substance dissolved in the water?

2 A sample of water gave a white residue when it was evaporated to dryness. The water also gave a yellow flame in the flame test and when tested with silver nitrate solution gave a yellow precipitate. After distillation no coloured flame was given in the flame test and no precipitate was given in the silver nitrate test.

    What conclusions can you draw from these results?

1 a What is the difference between a finite and a renewable resource?

  b Ethanol can be obtained from sugar by fermentation and from the reaction between ethene and steam. Explain which of these two processes is more sustainable.

2 The diagram below shows the main steps in the treatment of water to give potable water.

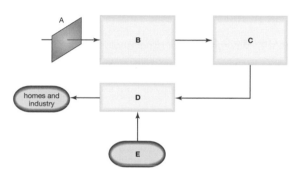

  a i Describe what is happening at A.

   ii What is added at C and why is it added?

   iii Explain why the water at D is not fit to drink.

   iv Describe the process taking place at E and explain why it is important.

  b Give two ways by which potable water is obtained from seawater.

  c Potable water is not pure water.

   i How could you show that potable water contains dissolved impurities?

   ii What physical property could you test to show that the water is not pure?

  d When waste water is treated the sludge formed after settlement is digested anaerobically.

   a What does anaerobically mean?

   b Give two useful products from anaerobic digestion.

H 3 a Explain what is meant by the following three terms when applied to the extraction of copper:

   i smelting   ii phytomining

   iii bioleaching.

  b You are given a solution of copper(II) sulfate. Give two ways you would obtain pure copper from the solution.

4 a What is meant by a life cycle assessment?

  b What are the four stages in the product's lifetime that are analysed for their impact on the environment?

5 The diagram to the right shows the arrangement of polymer chains in a thermosetting polymer.

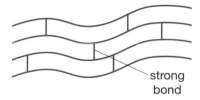

  a Explain why this type of polymer does not melt.

  b Explain why this type of polymer is a good choice for making electrical plugs.

6 a Give the equation for the formation of ammonia in the Haber process.

  b What are the conditions used in the Haber process?

  c Explain why a high pressure is used in the Haber process.

  d Give three compounds that would be found in an NPK fertiliser.

Glossary/Index

I

Incomplete combustion This takes place when a carbon compound burns in an insufficient supply of oxygen resulting in the formation of carbon monoxide or carbon particulates and water. **112**

Intermolecular forces The weak forces of attraction between molecules. **36, 38, 40, 110**

Ionic bond The electrostatic attraction between oppositely charged ions in a giant ionic structure. **32–5**

Ionic equation An equation describes a chemical reaction by showing the reacting ions only. **70**

Isotopes Atoms of the same element with the same number of protons and same atomic number, but with different numbers of neutrons and therefore different mass numbers. **15–16**

L

Law of mass conservation This states that matter cannot be created or destroyed, it means that a chemical equation must be balanced. **45, 47**

Le Chatelier's principle This states that when changes are made to a chemical system at equilibrium then the equilibrium will shift so as to cancel out the effects of the change. **105**

Life cycle assessment (LCA) An analysis of the environmental impact of a product from its production to its disposal. **151–2**

Limiting reactant The substance that is completely used up when a reaction is complete. **53**

Line spectrum When a coloured flame is analysed using a spectroscope to give a spectrum a series of lines are obtained. **132**

M

Malleable A substance is malleable if it can be hammered into shape. **21, 41, 42**

Mass number The total number of neutrons and protons in an atom. **13, 15**

Metal salt A compound formed when the hydrogen in an acid is replaced by metal. It has a metal part and an acid part. **26, 62, 67, 134**

Metallic bond The mutual attraction between delocalised electrons and metal cations in a giant metal lattice. **41**

Metals Elements that form positive ions. Metals tend to be shiny, malleable, ductile and good conductors of heat and electricity. **21–3**

Mixture Two or more different substances, elements or compounds mixed but not chemically join together. **9–11, 126, 128**

Mobile phase In paper chromatography the mobile phase is the solvent. In gas chromatography it is the carrying gas. **128**

Molar volume The volume of 1 mol of a gas. The conditions of temperature and pressure have to be stated when quoting this value. **58–9**

Mole A mole (abbreviation is mol) is the amount of a substance that contains 6.023×10^{23} (Avogadro's number) particles of that substance. **49–53, 56–9**

Monomers Small molecules that can be joined to other small molecules to form a long chain. In addition polymers the monomers are alkenes. In some condensation polymers the monomers are di-alcohols and di-carboxylic acids. **120, 122, 124**

N

Nanoparticles Very small particles with a diameter between 1-100 nm. Note: a nanometre is equal to 1×10^{-9} m. **43**

Negative ions or and ions Ions formed by the gaining of electrons, they move towards the anode in electrolysis. **14, 21, 32, 66, 68, 70, 134**

Neutralisation A reaction between an acid and an base or alkali to form a neutral solution of pH 7. **56, 66, 68, 71, 80**

Neutron A subatomic particle found in the nucleus of the atom. A neutron has no charge and has a relative mass equal to that of a proton. **13, 15**

Noble gas arrangements These are the electron arrangements of the noble gases. They are important because they are stable electron arrangements. **17, 19, 36, 140**

Non-metals Apart from the noble gases these elements tend to form negative ions and have properties which are different to metals. For example, they are poor conductors of heat and electricity and solid non-metals are brittle. **21, 24–7, 32**

NPK fertilisers These are fertilisers that contain nitrogen, phosphorus and potassium. **166**

Nucleus The centre of the atom containing the protons and neutrons (the nucleons). **12, 13, 15**

O

Ore A rock that contains enough of a metal to make it worthwhile to use it as a source of that metal. **64, 153**

Oxidation The gain of oxygen or the loss of electrons in a reaction. **65, 70, 73, 90**

Oxides of nitrogen These gases are formed when oxygen reacts with nitrogen at the high temperatures of a car engine, car exhaust or during thunderstorms. They cause respiratory problems and when dissolved in water form acid rain. **146**

Oxidised A substance is oxidised when it gains oxygen or loses electrons.

P

Particulates Small particles of carbon which cause global dimming (reduction of sunlight) and respiratory problems. **146, 147**

Percentage yield Percentage yield $= \dfrac{actual\ yield}{theoretical\ yield} \times 100\%$ **60**

Period The horizontal rows of elements in the periodic table. As you go across a period an electron shell is filled up. The period number of an element is the number of its occupied electron shells. **17**

Periodic table A table of the elements arranged in order of their atomic number. Elements with similar properties are in vertical columns called groups. **17–29**

pH scale A scale which runs from 0 to 14 and shows the acidity (value less than 7) or alkalinity (value greater than 7). pH 7 is neutral. **71, 118**

Photosynthesis The light-driven process by which plants and photosynthetic bacteria produce glucose and oxygen from carbon dioxide and water. **140**

Physical properties These properties can be measured or observed without changing the composition of a substance. **9, 21, 108**

Phytoextraction Plants take up a metal from slag or waste heaps that contain low-grade ores. The metal can then be extracted from the plants. **153**

Pipette This is used to measure volumes of liquids accurately. It is usually used in titrations where accurate volumes of solution need to be measured out several times. **80, 81**

Polymers Long chains of carbon atoms in a repeating pattern formed from smaller molecules called monomers. **40, 114, 120, 162–3**

Positive ions or cations These are ions formed when atoms lose electrons. They migrate towards the cathode in electrolysis. **14, 21, 32, 65, 70**

Potable water Water that is safe to drink. **155**

Products The substances being formed in a chemical reaction. These are written on the right-hand side of a chemical equation. **45, 47, 51–3, 60**

Answers

Atomic structure and the periodic table
Review it!

1   a $2Na(s) + Cl_2(g) \rightarrow 2NaCl(s)$

   b   i   Both have only 1 chemical symbol and therefore only 1 type of atom.

      ii   Na is a shiny silver white solid; Cl is a pale green gas.

      iii   Ions

   c   Because the sodium chloride is not chemically combined with the water and they can be easily separated by physical means.

2   The group is a vertical column of elements and a period is a horizontal row

3   Each element can only have one atomic number and that number is unique to that element. If it had an atomic number of 12 it would not be sodium.

4   Group 6, period 3.

5   a   X is found in the middle of the periodic table in the transition elements.

   b   Shiny; good electrical conductor; good thermal conductor; malleable; ductile; denser than the group 1 elements.

6   a   −1

   b   When they react the Group 7 elements gain one electron to form a stable outer shell, their reactivity depends on their ability to gain this extra electron. As the group descends, the outer shell is further from the positively charged nucleus and is shielded from the nucleus by an increasing number of electrons. This means that as you go down the group the attractive force on an electron being gained gets less and it gets harder to capture the extra electron.

7   a   It has 12 protons and 12 electrons.

   b   These are isotopes. Each isotope has the same number of protons but a different number of neutrons.

   c   Let there be 100 atoms of gallium. 60 atoms have a mass number 69 with a total mass of 4140 amu. 40 atoms have a mass number of 71 the total mass of 2840 amu.

      Therefore 100 atoms have a total mass of 4140 + 2840 amu = 6980 amu. The relative atomic mass is the average mass of each atom = $\frac{6980}{100}$ = 69.8 amu

8   Elements with similar properties were placed in vertical columns and ordered by their relative atomic masses. Where the known elements did not fit the pattern he left spaces for elements which had not yet been discovered.

9   Group 1 elements lose their outer electron when they react. As the group is descended, this outer electron is further from the attractive force of the nucleus and there are more shielding electrons between the nucleus and the outer electron. This means that the outer electron feels less of an attractive force and is more easily lost therefore making the lower elements more reactive.

10   The noble gases have stable full outer electron shells. This means they do not have to gain or lose electrons to become stable.

11   a   Mendeleev's periodic table was organised in groups of elements with similar properties. If argon had very distinct properties then it had to fit into its own group and therefore they had to be a group of elements with similar properties.

   b   Because it did not react with any other elements.

12   a   As the group is descended the elements get darker in appearance. Iodine is a dark grey solid; astatine is below iodine and would be darker in colour which suggests that it is black.

   b At_2

   c   NaAt

   d   -1. The ion is At^-.

Bonding, structure and the properties of matter
Review it!

1   The lithium atom loses its outer electron to form the Li^+ ion. The ion has a stable full outer electron shell.

2   a   1 nm to 100 nm (1 nm = 1×10^{-9} m)

   b   (Titanium dioxide nanoparticles in) sunscreens; (Silver nanoparticles are used in) antibacterial preparations; (Fullerenes are used to) deliver drugs and as lubricants.

3   a $\left[Mg \right]^{2+}$ $\left[\overset{\times}{\underset{\cdot}{Cl}} \right]^{-}$

   b $MgCl_2$

   c   The ionic bonds between the magnesium and chloride ions are very strong and because it is a giant structure all the bonds have to be broken. This requires lots of energy and a high melting point.

   d   In solids the ions are not free to move, therefore they cannot carry the current and do not conduct electricity.

4   a $H\overset{\displaystyle H}{\underset{\displaystyle H}{\overset{\times}{\underset{\times}{C}}}}H$

   b   Methane is a neutral molecule and the intermolecular forces between methane molecules are weak and require a small amount of energy to break them. Therefore methane has low melting and boiling points making it a gas at room temperature.

5   a   Giant covalent structure

   b   The bonds between the carbon atoms in both diamond and graphite are strong covalent bonds, all these bonds have to be broken. This requires lots of energy, so the melting point is high.

   c   The bonds in the layers of graphite are strong covalent bonds but between the layers the intermolecular forces are weak and easily broken allowing the layers to slide over each other easily.

   d   In graphite each carbon is bonded to other carbons leaving a spare electron. These spare electrons are delocalised in the layer and can carry an electric current making graphite a good electrical conductor. In diamond there are no spare electrons or charged particles making it a poor conductor.

6   A – Simple molecular   B – Giant ionic
   C – Giant metallic   D – Giant covalent

7   a   Methane has a simple molecular structure with weak intermolecular forces so it has low melting and boiling points. Potassium chloride has a giant ionic structure with strong ionic bonds between the ions. All these bonds need lots of energy to break them and therefore it has high melting and boiling points.

   b   The ions in magnesium oxide are Mg^{2+} and O^{2-}. In potassium chloride they are K^+ and Cl^-. The larger charges on the Mg^{2+} and O^{2-} means that their ionic bonds are stronger than those between the K^+ and Cl^- ions, these need more energy to break and therefore magnesium oxide has higher melting point.

   c   Zinc ions are larger than copper ions in the giant metallic lattice, this means that the layers of ions cannot slide over each other as easily, making the alloy a harder material.

   d   Sodium has a giant metallic structure in which there are delocalised electrons in both the solid and liquid states and these delocalised electrons can carry an electric current. This means that sodium is a good electrical conductor in both the solid and liquid states. Sodium chloride has a giant ionic structure. In the solid state the ions are not free to move and cannot carry an electric current so as a solid sodium chloride is a poor conductor. In the liquid state they can move and carry the current making sodium chloride a good electrical conductor.

Quantitative chemistry
Review it!

1   a   i 8.33×10^{-2};   ii 2.23×10^5
      iii 8.561×10^2 iv 4.53×10^{-5}

   b   i   4.00   ii 6.57×10^{-2}
      iii 4.55×10^{-2}
      iv 4.39×10^{-4} v 5.68×10^5

2   a   i $H_2(g) + Cl_2(g) \rightarrow 2HCl(g)$

      ii $2Na(s) + Br_2(l) \rightarrow 2NaBr(s)$

      iii $6K(s) + N_2(g) \rightarrow 2K_3N(s)$

      iv $Mg(s) + 2AgNO_3(aq) \rightarrow Mg(NO_3)_2(aq) + 2Ag(s)$

      v $4Na(s) + O_2(g) \rightarrow 2Na_2O(s)$

   b   The law of conservation of mass states that the mass of the reactants = mass of products; this means that the number and type of atoms on left-hand side of the equation must be the same as those on the right-hand side.

a 96 **b** 74 **c** 148 **d** 61

e 174 **f** 134.5 **g** 60

a 0.1

b $0.1 \times 6.02 \times 10^{23} = 6.02 \times 10^{22}$

c $0.1 \times 24dm^3 = 2.4dm^3$

a Atom economy method I = $\frac{44}{173} \times 100\% = 25.4\%$

Atom economy method II = $\frac{44}{44} \times 10\% = 100\%$

b Reduces waste

$HCl (aq) + NaOH(aq) \rightarrow NaCl(aq) + H_2O(l)$

Volume of NaOH = $\frac{30}{1000}$ = 0.03 dm³ Vol of HCl = 0.02 dm³

No. of moles of NaOH = C × V = 1 × 0.03 mol = 0.03 mol

No. of mol of HCl = no. of mol of NaOH = 0.03 mol

Concentration of HCl = $\frac{n}{V} = \frac{0.03}{0.02} = 1.5$ mol/dm³

a $\frac{6}{24}$ = 0.25 mol

b No. of mol of HCl = C × V = 1 × $\frac{200}{1000}$ = 0.2 mol

c From the equation 1 mol of magnesium reacts with 2 mol of HCl. Therefore 0.25 mol of magnesium react with 0.5 mol of HCl. There are only 0.2 mol of HCl and this is the limiting reactant.

hemical changes

view it!

a copper, iron, zinc, aluminium, magnesium

b i → zinc sulfate(aq) + copper(s)

   ii → NO REACTION

   iii → aluminium oxide(s) + iron

c i magnesium(s) + carbon dioxide(g) → magnesium oxide(s) + carbon(s)

   ii $2Mg(s) + CO_2(g) \rightarrow MgO(s) + C(s)$

   iii The magnesium gains oxygen – oxidation and the carbon dioxide loses oxygen – reduction

   iv I magnesium oxide II carbon

a The gas, hydrogen, is produced which can be tested for using a lighted splint. The gas pops.

b The copper is less reactive than hydrogen and will not displace it from the acid.

c Higher tier

   i **Zn(s) + 2H⁺ (aq) → H₂(g) + Zn²⁺ (aq)**

   ii **The zinc loses electrons when going from the neutral Zn to the positive Zn²⁺ ion. This is oxidation. The H⁺ ions gain electrons when forming H₂ and this is reduction.**

a Hydrogen at cathode; bromine at anode.

b A solution of potassium hydroxide

a Its solution is a weak alkali

b Phenol solution is a very weak acid

---

5 a i Dissolved in water – an aqueous solution

   ii NaOH(aq) + HCl(aq) → NaCl(aq) + H₂O(l)

   iii H⁺ (aq) + OH⁻(aq) → H₂O(l)

b i pipette ii burette iii conical flask

c The indicator changes colour.

Energy changes

Review it!

1 It decreases

2 A is an endothermic reaction because it is a thermal decomposition reaction.

B is an exothermic reaction because it gives out heat to warm up the food.

3 a The non-rechargeable cell will eventually stop producing a voltage because the chemicals will run out. The hydrogen fuel cell will keep on going as long as the hydrogen and oxygen are allowed to flow into the cell. Chemical cells can be used anywhere whilst the hydrogen fuel cell is hampered by the need for a supply of hydrogen to be available.

b The zinc is more reactive than the copper and therefore loses its electrons to form zinc ions more easily than the copper. This means that electrons would flow from the zinc to the copper making the zinc the negative electrode.

4 a i 23; 41; 7

   ii The units for temperature /°C.

b Use the same amount of acid (same volume/same concentration), the metals should have the same surface area (e.g. all 3 are powders), use either the same calorimeter/reaction vessel or identical ones.

c Thermometer or temperature datalogger, top-pan balance, well insulated calorimeter, spatula.

d Least reactive X, Z, Y Most reactive. X gives the lowest temperature rise then Z than Y which gives the greatest temperature rise.

5 a A Heat of reaction or energy change for reaction   B reactants

C Activation energy   D Products

E Energy   F Course of reaction

b Exothermic

c The energy required for the reaction to take place.

6 HIGHER TIER

a 4C—H, 1C══C, 1 Cl—Cl → 4C—H, 1C—C, 2C—Cl

The 4 C—H bonds are unchanged and can be omitted from the calculation.

1 × 610   1 × 245   1 × 350
2 × 345

The energy taken in to break bonds = 610 + 245 = 855 kJ

b **The energy given out when forming bonds = 350 + 690 = 1040 kJ**

c i **185 kJ/mol**

   ii **The energy given out is greater than the energy taken in, therefore the reaction is exothermic.**

Rates of reaction and equilibrium

Review it!

1 Measure the volume of carbon dioxide produced with time. Measure the loss in mass as time progresses.

2 a Measure the gradient of the tangent to the graph at any point.

b Rate = $\frac{25}{15}$ cm³/s = 1.67 cm³/s

3 a C because it has the steeper gradient at the beginning.

b C because increasing the temperature increases the rate.

c As the temperature increases the particles collide more frequently and with greater force, so the frequency of effective collisions increases and so does the rate.

4 a i A catalyst speeds up a chemical reaction and is unchanged chemically at the end of the reaction.

   ii 0.10 g

b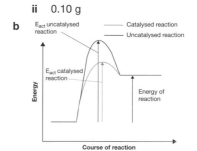

c A catalyst lowers the activation energy and this means that more particles have enough energy to react so that when they collide the collision is more likely to produce a reaction.

5 Temperature and surface area of a solid if one is involved. If two or more solutions are involved, then one concentration can be varied (the independent variable) whilst the others are kept constant.

6 a The ⇌ sign shows that the reaction can proceed both ways

b It is exothermic

c It turns blue and there is heat given out

7 Higher tier

a **The equilibrium will shift to the right.**

b **An increase in pressure favours the side with fewer gas molecules and this is the left-hand side.**

Organic chemistry

Review it!

1 a A = C₃H₈ B = C₃H₈O C = C₂H₄O₂ D = C₃H₆

b i A and D ii D iii C iv B v A

c A = propane B = propanol C = ethanoic acid D = propene

2 a The C═C group

b They have the same functional group and they have similar chemical properties.

c Add bromine water. The bromine water is decolourised.

d i

H—C—C—H with H, H on top, Br, Br on bottom

Answers

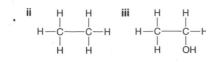

ii H-C(H)(H)-C(H)(H)-H **iii** H-C(H)(H)-C(H)(OH)-H

3 **a** Crude oil is composed of a mixture of miscible liquids with similar boiling points.

 b **i** The fractions become less viscous

 ii The boiling points decrease

 iii The fractions become easier to light

 c The fractions contain smaller molecules and this means that there are weaker intermolecular forces, therefore the molecules are more easily separated to become gases.

4 **a** C_nH_{2n+2}

 b **i** $CH_4(g) + 2O_2(g) \rightarrow CO_2(g) + 2H_2O(l)$

 ii $C_2H_6(g) + 3\frac{1}{2}O_2(g) \rightarrow 2CO_2(g) + 3H_2O(l)$

 c **i** A = cobalt chloride paper; B = limewater; C = to pump

 ii The cobalt chloride paper changes form blue to pink. This shows that water is formed.

 iii The limewater goes cloudy showing that carbon dioxide is formed.

5 **a** The amounts produced of the fractions with large molecules are more than required. At the same time the amounts of the fractions with smaller molecules are less than required. Cracking converts the larger molecules into smaller ones.

 b **i** C_6H_{14} **ii** C_9H_{20}

Chemical analysis
Review it!

1 **a** You need a nichrome wire, a Bunsen burner, a heat-proof mat, a watch glass and tongs. Place the solid being tested in the watch glass, add hydrochloric acid and then dip the nichrome wire into the solution/mixture. Hold the wire in a roaring blue Bunsen flame and note the colour formed.

 b **i** Yellow **ii** Green

 iii Orange-red **iv** Lilac

2 **a** Effervescence/fizzing

 b **i** Yellow precipitate

 ii White precipitate

 iii Cream precipitate

 iv White precipitate

 c **i** White precipitate

 ii No change/no reaction

 d **i** White precipitate

 ii White precipitate that re-dissolves on adding excess sodium hydroxide solution.

 iii Green precipitate

3 X = magnesium sulfate

 Y = sodium bromide

 Z = potassium chloride

Chemistry of the atmosphere
Review it!

1 Carbon dioxide, water vapour, methane, nitrogen and ammonia

2 Dissolving in the water forming the oceans. Uptake by plankton in the sea which then form their shells; they are compressed by sediments and form limestone. Photosynthesis by plants. Plankton covered by sediments in the absence of oxygen form oil. Plants covered by sediment then compressed form coal

3 78% nitrogen; 21% oxygen; the remaining 1% consists of noble gases and approximately 0.04% of carbon dioxide

4 Carbon dioxide and methane absorb infrared radiation and then are re-radiated back to Earth, warming up the atmosphere.

5 The recent rises in carbon dioxide levels are mirrored by increased temperature in the atmosphere.

6 The increased global temperature causes the ice caps to melt, thus increasing the water in the oceans and the water levels rise given rise to floods. Also severe storms lead to greater rainfall and flooding.

7 The total amount of carbon dioxide emitted over the lifetime of an activity or product.

8 • Energy conservation will reduce the amount of carbon dioxide produced by burning fossil fuels.

 • The use of alternative energy resources will also reduce the amount of carbon dioxide produced by burning fossil fuels.

 • In carbon capture and storage carbon dioxide produced in power stations is pumped into deep underground porous rocks at the sites of exhausted oil wells.

 • Carbon taxes penalised people/companies/organisations that use too much energy and this will inhibit people from overuse of energy.

 • Carbon offsetting is when plants are planted which taking carbon dioxide through photosynthesis thus reducing the amount of carbon dioxide in the atmosphere.

9 • People are reluctant to change their lifestyle. For example, they still use large cars which consume more energy.

 • Countries do not cooperate with each other.

 • Some countries still believe that global warming is a natural phenomenon and is not caused by humans.

 • People are still unsure of the facts and the consequences of global warming.

 • Countries still find it economical to use fossil fuels

10 An atmospheric pollutant is something that is introduced into the atmosphere and has undesired or unwanted effects.

11 **a** **i** CO

 ii By the incomplete combustion of carbon-containing fuels.

 iii It is toxic because it reduces the amount of oxygen getting to the brain.

 b **i** Oxides of nitrogen are formed by the reaction between nitrogen and oxygen at high temperatures.

 ii For example, in car engines and exhausts and in thunderstorms.

12 Sulfur dioxide is formed by the reaction of sulfur and oxygen. It dissolves in water to form acid rain and it causes respiratory problems.

Using resources
Review it!

1 **a** A finite resource will run out but a renewable one can be replaced.

 b The fermentation of sugar is the sustainable process because the sugar can be regrown again whilst th ethene comes from the cracking of crude oil which is a finite resource.

2 **a** **i** Large objects are screened out o the water

 ii Aluminium sulfate is added to ma small particles clump together and settle to the bottom of the tank

 iii It contains bacteria

 iv Chlorine is added OR the water i treated with UV light to kill bacter

 b Distillation or reverse osmosis

 c **i** Evaporate off the water

 ii The boiling point

 d **i** In the absence of oxygen

 ii Methane and fertilisers

3 **a** **i** Roasting with carbon

 ii Plants which absorb copper are planted on sites where there are low-grade copper ores. After they have grown they are harvested and burned to leave copper deposits in the ashes.

 iii Bacteria use low grade copper sulfide ore in heaps for an energy source and they oxidise the ores The liquids leaching from the heaps contain copper ions.

 b Electrolysis and displacement of the copper by adding a more reactive metal such as iron.

4 **a** A life cycle assessment is an analysis the environmental impact of a product at each stage of its lifetime from its production all the way to its disposal.

 b The extraction/production of raw materials. The production process – making the product, including packaging and labelling. How the product is used and how many time it is used. The end of the life of the product – how is it disposed of at the end of its lifetime. Is it recycled?

5 **a** The chains cannot slide over each other because they are joined by strong covalent bonds.

 b It will not melt if the plug gets hot an it does not conduct electricity and is therefore safe to handle.

6 **a** $N_2(g) + 3H_2(g) \rightarrow 2NH_3(g)$

 b A temperature of 450°C; a pressure of 200-250 atmospheres and an iron catalyst.

 c The formation of ammonia means that the number of gas molecules goes down from 4 to 2. If the pressu is raised the chemical system will try and lower it by making fewer gas molecules and this means that formation of ammonia is favoured by a high pressure.

 d Ammonium nitrate, ammonium hydrogen phosphate and potassium chloride.